Cutting the Cost of Cold

Cutting the Cost of Cold

Affordable warmth for healthier homes

Edited by

JANET RUDGE & FERGUS NICOL

London and New York

First published 2000 by E & FN Spon
11 New Fetter Lane, London EC4P 4EE

Simultaneously published in the USA and Canada
by E & FN Spon
29 West 35th Street, New York, NY 10001

E & FN Spon is an imprint of the Taylor & Francis Group

Printed and bound in Great Britain by
TJ International Ltd, Padstow, Cornwall

Publisher's note
This book has been produced from camera-ready copy supplied by the editors.

British Library Cataloguing in Publication Data
A catalogue record for this book is available from the British Library

Library of Congress Cataloging in Publication Data
Cutting the cost of cold: affordable warmth for healthier homes / edited by Janet Rudge
and Fergus Nicol.
    p. cm.
  Includes bibliographical references and index.
  ISBN 0-419-25050-6
    1. Housing and health—Great Britain. 2. Dwellings—Heating and ventilation—Health
aspects—Great Britain. 3. Mortality—Environmental aspects—Great Britain. I. Rudge,
Janet, 1949– II. Nicol, F. (Fergus)

RA616.C88 2000
613′.5—dc21   00-021310

Editors' note and acknowledgements

There is increasing interest in the relationship between health, buildings and energy use. Brenda Boardman points out in her Introduction that the links have now been officially recognised. However, one barrier to effective action to cut the ill health associated with cold, damp dwellings is the number of different Government Departments, authorities and professions involved in its delivery.

Our own research interest in the field is from a buildings and energy use background. In February 1999 we organised a symposium on Health, Housing and Affordable Warmth at the University of East London School of Architecture. Our objective was to bring together academics and professionals from the varied but relevant disciplines for a useful exchange of information on current research and practice. The success of this meeting suggested to us that a book was needed. Fuel poverty and inadequate housing need to be tackled on many fronts and, by taking a multi-disciplinary approach, we hope this book will be of use to practitioners as well as researchers in all the related fields, whether they be policy, housing, health, anti-poverty or building.

We have arranged the book into four parts which are introduced by Brenda Boardman. The first Part describes current research bearing on cold, ill-health and energy use. Part Two deals with tools for identifying and costing the problems associated with fuel poverty and cold damp housing. In Part Three, case studies and examples of practice are described where partnerships have been formed to implement affordable warmth and remedy illness. The benefits and difficulties of partnership working are discussed as a model for future action. Part Four brings together proposals for the future from three different standpoints – those of housing studies, health services and sustainable architecture.

We would like to record our thanks to a number of people: firstly to the authors who put in the hard work and have been very patient with our demands; secondly to John Appleby, Trish Brady, David Crowther, Liddy Goyder, Mark McCarthy, Ravi Mahreswaran, Richard Moore and Chris Saunders and the EAGA Trustees, who all helped with the chapter reviewing process; thirdly to Peter Salter, of UEL for letting us stage the symposium and to EAGA-CT and Newham Council for giving it their endorsement; lastly to Mike Doggwiler and Edith Henry of E&FN Spon and to their former colleague Tim Robinson who commissioned the book.

Janet Rudge and Fergus Nicol,
LEARN, University of North London, November 1999

Contents

Part Three: Inter-agency partnership in practice

Part Four: Ways forward

Abbreviations

Organisations

ACE	Association for the Conservation of Energy
BRE	Building Research Establishment
DETR/DoE	Department of the Environment, Transport and the Regions formerly, Department of the Environment
DoH	Department of Health
EAGA - CT	Energy Action Grants Agency – Charitable Trust
ECSC	Energy Conservation and Solar Centre
EST	Energy Saving Trust
NEA	National Energy Action
NES	National Energy Services Limited
ONS/OPCS	Office of National Statistics formerly Office of Population, Censuses and Surveys

Common acronyms

HAZ	Health Action Zone
BREDEM	Building Research Establishment Domestic Energy Model
HIP	Housing Investment Programme
HEES	Home Energy Efficiency Scheme
HECA	Home Energy Conservation Act
AWI	Affordable Warmth Index
NHER	National Home Energy Rating
SAP	Standard Assessment Procedure
SRB	Single Regeneration Budget

About the authors

Professor Peter Ambrose taught at Sussex University for over thirty years and was the Director of the Centre for Urban and Regional Research. He is now retired and is Visiting Professor in Housing Studies at Brighton University. He is currently engaged on the 'Benchmarking' for the East Brighton New Deal for Communities programme. He is the author of seven books and numerous other publications.

Dr Martin Bardsley is currently the manager of the Health of Londoners Project, which produced the first Public Health Report for London, which examined a wide range of issues concerning health in the capital. His previous experience includes work on health needs assessment and the development of outcome measures for Regional Health Authorities.

Dr Brenda Boardman MBE researches into the efficient use of energy at the University of Oxford. She is Head of the Energy and Environment Programme at the Environmental Change Unit. Dr Boardman's main research focus is on the way that energy is used in British homes, particularly by low-income households – the problem of fuel poverty. She considers the economic, social and technical aspects of the subject and has a strong policy perspective to her work.

Professor John Brazier is Professor of Health Economics and director of the Sheffield Health Economics Group at the School of Health and Related Research at The University of Sheffield. He has researched and published widely on the measurement health outcomes for a range of medical and population interventions.

Professor Jake Chapman has worked at the Open University since 1970, initially as a lecturer in Physics, then as Director of the Energy Research group and, since 1979, Professor of Energy Systems. A co-founder of Energy Advisory Services Ltd, he developed the Milton Keynes Energy Cost Index which laid the foundations for the NHER and SAP energy ratings. Professor Chapman was Managing Director of NEA until he retired in 1998. In 1995 he received a Royal Society Gold Medal and Esso Award for his contributions to energy conservation.

Dr Ken Collins MB BSc DPhil MRCS MRCP was, until retiring recently, a Senior Clinical Lecturer at University College Hospitals, London and a Member of Staff of the Medical Research Council. He has published widely in scientific journals on human physiological responses to the temperature environment, particularly on the effects of extremes of heat and cold, and on thermo-regulation in the elderly. He has served on WHO Scientific Committees concerned with housing conditions and health and the effects of climate change. He is the author of a number of books.

Roger Critchley is an environmental health consultant and building surveyor and has nearly 30 years experience of improving housing conditions for low-income households. He has undertaken consultancy work for EAGA and EAGA Charitable Trust, and had been involved in a number of complex and successful court actions involving energy audits, cold homes and low-income households. He is a member of the Health and Housing Group whose main work involves inspecting housing rented by low-income households.

Dr Gavin Donaldson obtained his BSc at North-East London Polytechnic in 1984 in Biophysical Science and a PhD in 1989 at London University. He has worked at the Royal London Hospital and later Queen Mary Westfield College for 15 years undertaking both laboratory and epidemiological studies on the causes of increased mortality and morbidity during hot and cold weather.

Jan Gilbertson is a Research Associate in the Centre for Regional, Economic and Social Research at Sheffield Hallam University. She has is responsible for survey management and analysis on a number of large-scale housing investment and health projects.

Dr James Goodwin is a lecturer in Medical Science at St Loyes School of Health Studies and research scientist in the School of Postgraduate Medicine and Health Sciences, University of Exeter. He recently completed a PhD at the University of Exeter into problems of winter cold in the elderly. Has conducted numerous investigations with Dr Ken Collins.

Dr Geoff Green is a Senior Research Fellow in the Centre for Economic and Social Research at Sheffield Hallam University and advisor to the European Regional Office of the World Health Organisation. He co-ordinates a number of research studies investigating the impact of housing investment on the health and quality of life of residents.

Rob Howard is a Senior Projects Officer with NEA, specifically responsible for supporting and developing NEA's special projects in the East Midlands. For nearly five years he has been coordinating the work of the Nottingham Project, which has worked closely with Nottingham City Council in helping to develop its energy efficiency strategies.

Dr Stirling Howieson is currently Director of Studies in Building Design Engineering at the University of Strathclyde. He was formally General manager of Community Architecture Scotland and the Technical Services Agency, both community based organisations active in the field of housing and research. He is author of *'Housing - Raising the Scottish Standard'* which set out a methodology for determining if a dwelling is "thermally safe".

Adrian Jones MCIEH qualified as an Environmental Health Officer in 1978 and has recently worked on primary care initiatives and on housing improvement schemes, which have aimed to regenerate the inner city of Birmingham. This has included development work with communities living in housing renewal areas and health action areas. Adrian leads a team developing the Urban Care Strategy, an alternative approach to property maintenance and healthy homes in which community self-help groups manage part of the urban renewal process.

Emma Jones has a BA (Hons) in geography from Cambridge University and an MSc in environmental technology (Energy policy) from Imperial College. She joined the Association for the Conservation of Energy in 1997 as a researcher, working primarily on local authority projects, including the Energy Saving Trust's 'HECAction' awards scheme for authorities. In June 1999 she joined the Improvement and Development Agency for Local Government as an Energy Consultant.

Professor William Richard Keatinge MA,MB, BChir, PhD, FRCP studied problems of survival in cold water after shipwrecks as a Navy doctor in one of the last national service intakes. His long career has always returned to health problems caused by heat and cold. He was Co-ordinator of the European Union's Eurowinter and Russian Winter research programmes on cold-related deaths and protection against cold in the varied climates of Europe and Russia.

Lutfa Khanom is working as a researcher for the Tower Hamlets Health Strategy Group. Her background is in Social Policy Research. Initially working to help local youth among different ethnic minority communities, she progressed to working for people of different age groups. Most of her work has been centred in the borough of Tower Hamlets.

Megan Landon is a Research Fellow in the Environmental Epidemiology Unit at the London School of Hygiene and Tropical Medicine. She graduated in Social Anthropology in 1987 and has an MSc in Environmental Epidemiology and Policy. Her main research interests are in the relationship between housing and health, climatic influences on health and environmental health risk assessment.

Alan Lawson completed a B.Eng. (Hons) degree in Building Design Engineering at the University of Strathclyde, specialising in environmental engineering. He then undertook an MSc in Energy Systems and the Environment before joining CEDAR (Centre for Environmental Design and Research) within the Architecture department, where he is currently researching the relationship between asthma and the built environment.

Dr Ian Mackenzie was a general practitioner for ten years before becoming an FHSA Medical Adviser and training in public health medicine. He is Chairman of the Cornwall Asthma Task Force and has research interests in asthma and primary care. He has recently been seconded as Director of Developments to the Health Authority board.

Fergus Nicol is a building physicist whose main field of research over a number of years has been indoor thermal comfort. He is currently a senior researcher at the Universities of Oxford Brookes and North London, and is coordinator of a major European thermal comfort project.

Professor Tadj Oreszczyn PhD, CEng, MCIBSE, MInstE trained as a physicist at Brunel University and then undertook a PhD in solar energy at the Open University. He then joined Energy Conscious Design as a Senior Energy Consultant before joining the Research in Building Group at the Polytechnic of Central London. At the Bartlett School of Architecture he has been Course Director of the Masters in Environmental Design and Engineering and Director of the Energy Design Advice Scheme. He was appointed as Professor of Energy and Environment in 1999.

Dr David Ormandy is principal research fellow at the School of Law, University of Warwick. He is currently responsible to the Department of the Environment, Transport and the Regions for the development of a Housing Hazard Rating System.

Dr Jean Peters graduated from Loughborough University of Technology. She joined the Sheffield University Department of Public Health in the School of Health and Related Research (ScHARR) in 1994 as a Research Fellow. Her current interests include the impact of environmental factors such as air pollution, and lifestyle factors such as smoking, on health, with a special focus on respiratory health. Other interests include the use and interpretation of outcome measures for chronic diseases, such as asthma and diabetes.

Helena Poldervaart is a Project Director at Projects in Partnership, a charity working on sustainable development, particularly in the field of consensus building and partnership working. She has been involved in projects on energy efficiency for 5 years, bringing together interested parties such as industry and householders to identify solutions and then developing and testing them.

Stephen Pretlove BSc(Hons) MSc(Arch) MCIOB MBEng is a Senior Lecturer in Building Science in the Faculty of the Built Environment at South Bank University. He has a degree in Building Engineering and a Masters degree in Architecture (Environmental Design and Engineering). His PhD involves the development of a computer model that predicts the internal environmental conditions in dwellings in order to assess moisture related health risks.

Janet Rudge BA(Hons) BArch MSc, Registered Architect, has worked as an architect in a variety of practices, both in the public and private sectors. Much of her experience has been with housing associations involved with rehabilitation of dwellings in northeast England. Since 1994 she has worked as a teacher of Environmental and Energy Studies with both undergraduates and MSc students. She is currently researching housing, energy efficiency and health for a PhD at the University of North London.

Brian Scannell has worked within the environmental field since his graduation in 1982. In 1985 he joined Welsh Water and following privatisation in 1989 he helped to establish an environmental consultancy business. In 1994 he moved to Hong Kong as Associate Director of an environmental consultancy business. He joined National Energy Services Ltd in 1997 as Sales and Marketing Director and became Managing Director in 1998.

Professor Peter F Smith is Professor of Architecture at Sheffield Hallam University. He gained his architectural education at Cambridge University and received his PhD from Manchester University. From 1992 he has conducted research within the environmnetal field for the National Audit Office and the European Commission. He is currently conducting research for the Engineering and Physical Sciences Research Council into the upgrading of the national housing stock. Between 1987 - 96 he was chairman of the RIBA Environment and Energy Committee and is now chairman of the RIBA Sustainable Futures Committee.

Dr Margaret Somerville qualified in medicine from Newcastle-upon-Tyne in 1978 and initially pursued a career in general and respiratory medicine.

Following a period of research in respiratory physiology, she specialised in public health medicine. She has held her current position, as consultant in public health medicine with South and West Devon Health Authority, for the last 4 years.

Dr Matt Stevenson graduated from the Operational Research Department of Lancaster University in 1997. He works for the School of Health and Related Research (ScHARR) at Sheffield University, where he has undertaken modelling with a primary focus in disease pathway modelling.

Simon Stevenson is a research fellow in the Environmental Epidemiology Unit, London School of Hygiene & Tropical Medicine. His main research interests are in the use of small area spatial analysis for investigating environmental influences on health. His current areas of work are in seasonal variations in mortality and air pollution epidemiology.

Helen Thompson gained a Social Science degree at Middlesex Polytechnic in 1985 and carried out multi-sector work in the voluntary sector from 1985 to 1998. She became Health Action Group Officer at Nottingham Health Authority in June 1998 and Public Health Development Worker with the Directorate of Public Health in August 1999.

Dr Paul Wilkinson graduated in clinical medicine from Oxford University in 1985, and joined the Epidemiological Research Unit at the National Heart and Lung Institute. He is now Senior Lecturer in Epidemiology in the Environmental Epidemiology Unit, London School of Hygiene and Tropical Medicine. His main current research interests are air pollution and climatic influences on health and the use of small-area geographical methods in epidemiological research of environmental hazards to health.

Victoria Wiltshire is a Senior Researcher at the Association for the Conservation of Energy. She is a member of the HECAction team which monitors the schemes funded under the programme and is also on the steering committee for the National Right to Fuel Campaign and a member of the Fuel Poverty Strategy Group of the Energy Efficiency Partnership for Homes (EEP). Her work primarily involves the social aspects of energy efficiency such as employment and health issues.

Cutting the cost of cold

Affordable warmth for healthier homes

Introduction

Introduction

Brenda Boardman

This book is timely for two reasons: first, the Government has acknowledged that fuel poverty exists and has set the objective of eradicating it (the former Minister for Energy Efficiency, Angela Eagle, in June 1998). Secondly, several official documents have confirmed the link between cold, damp housing and ill health [eg Acheson, 1998]. The most recent, the consultation document on the Home Energy Efficiency Scheme - *The New HEES* [DETR 1999], targets the most vulnerable to ensure that energy efficiency improvements are reducing their health risks. These are both major political steps forward in the task of providing affordable warmth to UK residents.

Now that the problem of fuel poverty is accepted, the focus is on the task of defining our objectives and the ways to make progress. For those of us who have been campaigning to get recognition of fuel poverty - and many of them are contributing to this book - this is a welcome, but challenging situation. Now we have to come up with solutions, not just campaigning statistics.

Fuel poverty was defined as the inability to obtain adequate energy services for 10% of income [Boardman, 1991, p201]. That is still the accepted definition, but greater clarity is now required about the components of 'income' and 'adequate energy services'. For instance, does income mean disposable income, after the exclusion of all housing costs? Is it just the money that goes through the household purse? As housing benefit often goes directly to the landlord and is never seen by the householder, the fairest definition of disposable income would exclude all housing costs. Is it accepted that the appropriate heating regime is 21 °C in the living room and 18 °C elsewhere? This is the standard definition used in the English House Condition Survey [DoE, 1996, p82]. The poor, after all, are rarely out of the house as they are typically pensioners, single parent families, the sick, disabled and unemployed and need these temperatures for 16 hours a day.

The 10% has to cover all energy costs, not just those for heating, despite the phrase 'affordable warmth'. Affordable warmth is the opposite of fuel poverty. As the originator of this expression, I have to plead guilty to causing some confusion and implying that it is only warmth that we are concerned about. Heating is the most important use of energy in the home (about 60% of energy expenditure) and of course is the way to ensure that people stay comfortable and healthy. Affordable warmth is also a good, positive, catchy phrase that has achieved widespread support in the last ten years. However, it is really an abbreviated form of saying that an adequate standard of all energy services should be affordable.

At the time of writing (October 1999), the response to the HEES consultation document has not been issued, so that it is not clear what definition is being proposed for an absence of fuel poverty, i.e. the availability of affordable warmth. The suggestion here is that it is: *adequate energy services for 10% of income (excluding all housing costs)*. This is the definition used in the Affordable Warmth Index, described by Chapman and Scannell.

Just as there is now a debate about the precise definition of fuel poverty, so there is a continuing discussion on the causal links between poor quality housing and bad health. The inescapable fact is that the UK has a higher increase of deaths in winter over the summer level than almost any other European country - Ireland is similar. The rate of Excess Winter Deaths (EWD) has been dropping in Britain, largely because we are having fewer cold winters, but also because of the increasing presence and use of central heating. However, the UK is still exceptional and the implication must be that the problem is the low level of energy efficiency of our housing stock. Even much colder countries, such as Russia, do not experience such a high winter mortality peak, because they have warmer homes. These winter deaths are not from hypothermia - a common misconception. That is the cause of death for only a few hundred people each year, out of the 20-40,000 deaths. The main causes of cold-related ill health and death are linked to respiratory and circulatory problems. There is, for instance, no seasonality to cancer deaths.

As the chapters in this book expose, there is not clear agreement on the causal links between environmental conditions and subsequent ill health. For instance, the balance between respiratory and circulatory (cardiovascular) disease may depend on the area (Greater London vs. UK), the age group (elderly vs. whole population) and the period you choose (winter can be three, four or six months long in different statistical analyses). Whilst it is preferable to understand the processes that are causing these excess winter deaths and, ultimately, the way to prevent them, the acceptance of a general link means that action can be taken to improve the energy efficiency of dwellings and improve the health of the present and future residents. Whether a warm, cheap-to-heat, energy efficient home provides healthy living conditions because it is warm, free of mould or no longer damp, is not as important as making sure that more people have the choice of living in good quality housing.

Some of the uncertainties come from combining three sets of problematic parameters and three sets of disciplines:

- ignorance about the causes of disease;
- variability of housing conditions;
- the wide range of parameters in human lifestyles.

For instance, with asthma what is the relative effect of mould, dust mites and passive smoking? And how do you isolate these factors? Levels of income, clothing and nutrition may all be similarly inter-twined. It is easier to find associations, than to establish causality and, to a certain extent, this is sufficient. The evidence for better housing is compelling.

One of the values of this book is that it brings together papers by authors from each of these perspectives and provides the evidence in a form that is accessible to people from the other disciplines. This is an inter-disciplinary problem and the various professionals need to understand each other's work, so that they can all collaborate more easily: a real challenge.

There are different ways of providing good quality homes and several are mentioned in the case studies in this book. The landlord, usually a social housing provider, can fund the improvements and, increasingly, the health authority can contribute. Communities themselves can, through self-help and support, prevent small problems becoming large ones. The breadth of experimentation with processes, monitoring and evaluation tools is demonstrated through these pages. Often the work is on-going, but is described here to encourage others to be innovative and implement their own ideas. The problem of fuel poverty is vast. Estimates vary from a minimum of 4.4m homes in England, based on the 1999 HEES consultation document, and possibly as many as 8m in the UK, using Angela Eagle's figures in 1998. In all cases, multiple initiatives are needed, now.

CURRENT RESEARCH

As Professor Keatinge and Gavin Donaldson state in Chapter 1, the debate about excess winter deaths is a debate about the extent this is as a result of getting cold indoors and getting cold outside. As the originator of the 'getting cold at bus stops' school, it is interesting that Professor Keatinge now believes, based on five European countries, that both cold homes and inadequate clothing when outside are equal contributors. People are either good at insulating themselves and their homes, or they are not. They either take winter cold as a serious threat to health, or they do not. The British, it seems, do not, hence our high winter mortality rates. Whatever the causes - and the coronary and respiratory temporal patterns and medical evidence are summarised - there is much that can be done. 'Adequate protection against cold can virtually eliminate excess winter mortality' and this includes both improved energy efficiency of the housing stock as well as telling people to wear woolly hats.

A useful resume of the epidemiological debate is given by Wilkinson, Landon and Stevenson. They believe that cold indoor temperatures account for the 'greater part' of the seasonal increase in deaths. They give it greater importance than Keatinge, which demonstrates the problems of precision in seeking the causes. As this is written from a medical perspective, they state that there is little 'direct evidence' of the effect of housing condition. This statement is challenged by later chapters, such as Green *et al.*. However, the presence of central heating is one of the simple indicators associated with a decreased risk of winter deaths.

Wilkinson *et al.* confirm, graphically, the 40% increase in winter deaths (between peak winter and lowest summer) using 11 years of recent data from London. The disadvantage is still there and still large. These deaths are not caused by a sharp effect from the coldest temperatures, but are a consistent, gradual response to deteriorating weather conditions as winter approaches.

Dr Ken Collins is another doyen of the fuel poverty campaigners and gives an excellent summary of the medical evidence on respiratory illness. It is widely accepted that cold, damp housing is unhealthy, but the relative effects of cold, damp and mouldy living conditions are difficult to disentangle as they are co-related and, consequently, the different importance of mould and dust mites in causing asthma cannot be identified.

He gives evidence that there is a greater increase in winter mortality from respiratory disease than from circulatory (coronary), and that respiratory health is more related to indoor temperatures and cardiovascular to outdoor cold. These distinctions need to be examined further. Collins confirms that the elderly need warmer temperatures, because they are less active, but have colder homes. Methodologically, it is very difficult to show a definitive link between home temperatures and specific health outcomes, just as it is nearly impossible to identify the exact thermal conditions of the home after someone has died.

James Goodwin focuses on the elderly and provides a meticulous resume of the evidence related to coronary disease and confirms that:
- warm homes are good
- physical activity is good
- but the shock of cold morning might cause too much cardiovascular strain, particularly if leaving a cold dwelling. The effect is less if leaving a warm dwelling.

This last evidence may be the link between his work and Keatinge's: the effect of external cold is mediated by the extent to which you were warm indoors.

Blood pressure, which is an indicator of likely circulatory problems, such as heart attacks and strokes, peaks in winter for us all, but slightly more for elderly people. For the latter, the strongest relationship is with cold, indoor temperatures during the day. There is no seasonal variation in blood pressure at night and standard day-time room temperatures, across the year, remove the seasonal effect of increases in blood pressure.

The substantial growth of asthma, particularly amongst children and in the homes of people from the lower socio-economic groups, is being investigated by Howieson and Lawson in south-west Scotland. They are looking at ways to reduce the conditions that favour dust mites (through steam cleaning and extract fans) and whether changed conditions result in reduced populations and lower levels of asthma. As Ken Collins describes, dust mites are more commonly associated with warm, damp homes, whereas mould is with cold and damp. The latter is a greater indicator of fuel poverty, whereas the former appears to be worse for asthma.

The Howieson and Lawson study is therefore more to do with health and housing than with affordable warmth. This study of 70 asthmatics will be completed in March 2000 and will provide evidence on the relative effects of cold, mould, urea formaldehyde foam in the walls (which can also cause an allergic response) and dust mites.

Lutfa Khanom reports from Tower Hamlets, in London, on the impact of energy efficiency improvements on households suffering from fuel poverty and their health problems. This, the first stage of the study, examines the self-reported health of 30 public sector households (89 people interviewed) and examines their coping strategies. Over two-thirds reported suffering from depression or worry and this was significantly associated with, for instance, wallpaper peeling off in damp conditions. The residents of cold homes showed high levels of wheezing and nausea.

This study looks at the mental distress suffered, largely due to a feeling of helplessness, particularly in relation to the conditions their children have to live in. The low-income tenants are unable to pay higher fuel bills, cannot afford to decorate and are dependent upon the housing department for improvements.

Geoff Green, David Ormandy, John Brazier and Jan Gilbertson look at the effect of energy efficiency improvements in Sheffield tower blocks on warmth and comfort, damp and mould and therefore better health status for the residents. There are problems with both longitudinal studies (the same property before and after improvements) and when the comparison is between matched, blocks, some improved some not, as here. Most potential confounding factors were recognised and accounted for, but the findings are limited because of the higher proportion unemployed and the slightly larger families in unimproved properties.

Thus the very real benefits of higher temperatures (7.1°C warmer) and the better health of the residents in the improved blocks cannot be attributed categorically to the energy efficiency improvements. It could be linked also to the higher income levels of employed, smaller households. On all health dimensions, the residents of the improved blocks had higher mean scores, indicating better health, than the residents of the unimproved blocks. On six dimensions, the residents of the improved blocks, who are relatively deprived, reached a similar level of health to the Sheffield average. Good housing can compensate for deprivation. One message that Green *et al.* feel results from their research is the benefit of investing in capital improvements to the housing stock, rather than taking the simple, but recurring approach of increasing incomes. In the latter case, the underlying problem is never tackled.

TOOLS FOR RESEARCH AND PRACTICE

The next four papers are about tools: models and mapping, tried and in the process of being developed. As fuel poverty and the need for affordable warmth are accepted, there is a growing focus on methods of identification and monitoring or evaluation.

The Affordable Warmth Index, described by Jake Chapman and Brian Scannell, is based on a definition of affordable warmth where housing costs are excluded from the assessment of income. This is the most generous definition and the clearest.

The AWI is a useful, practical computerised method of identifying the extent of fuel poverty for a specific household. The AWI is being trialled in nine situations and is applied at the level of the individual house. Whilst the results are not known yet, this chapter gives a useful summary of auditing methods and the issues that need to be considered when assessing affordable warmth in the poorest households. The explanation provides a good primer for anyone thinking about affordable warmth. The first case study to be analysed demonstrates that the Standard Assessment Procedure (SAP) is not an appropriate alternative to the AWI for fuel poor households. The problems of the fuel poor are qualitatively different from those of other households.

Tadj Oreszczyn has developed an alternative approach - the mould index. Relative humidity is the key parameter for determining if mould or dust mites will survive. The parameters that affect the level of relative humidity are described and a computerised mould index, based on BREDEM 8, is introduced. By combining this with the affordable warmth index, it is possible to demonstrate that improved energy efficiency is the best solution for warm, mould-free homes. Attempting to cure the problem of mould solely by increasing ventilation rates does not provide affordable warmth. By combining these two approaches, it is possible to identify inappropriate strategies, for instance installing double-glazing in the absence of better insulation. Forced ventilation, with mechanical extraction, reduces moisture, but puts up the cost of keeping warm; it is, therefore, a more appropriate intervention for the fuel rich than for the fuel poor.

Janet Rudge is developing a mapping tool to provide the evidence for linking winter morbidity, low incomes and energy efficiency, using data from the London Borough of Newham and the local health authority. Using data already available, she overlays households on low incomes (probably a combination of housing benefit and council tax benefit), houses with a low level of energy efficiency (average SAP ratings for each ward based on HECA, EHCS, age of property, etc) and those who suffer from cold-related illnesses in winter. This is undertaken for enumeration districts or wards: there are 24 wards in Newham, 460 enumeration districts and a population of 220,000.

Both the health and local authorities hold considerable amounts of data, which can be combined effectively to target resources. However, persuading the authorities to allow the data to be used and combining the various datasets is a complex process. The paper provides a good discussion on several of the pitfalls. As Rudge states, many of these databases are in the process of being improved and used more regularly, so queries are useful in improving the standard of information. Once set up, the combined evidence provides the basis for long-term analysis, of maintenance costs as well as health benefits.

Under an agreement between the promoter of a private member's bill, John McAllion, and the Government, local authorities are now required to identify their plans for tackling fuel poverty and annual progress. Janet Rudge's mapping techniques could provide the basis for complying with the McAllion Option.

An alternative lever is demonstrated by Jean Peters as they identify, through modelling, the costs to the health service of cold damp housing. It is difficult to estimate the costs to the National Health Service - condensation-related ill health is thought to cost about £1,000m annually [Hunt and Boardman 1994, p30]. Peters looks at the savings that could be made and the assessment provides an indication of present costs and therefore the gains that would come from improved energy efficiency of the housing stock. The estimated annual saving is £44-113m: the equivalent of 0.2% of NHS expenditure in 1992/3. However, not all of these health costs can be saved immediately and, as the authors state, this could be a conservative estimate. The sums relate to morbidity only, not to mortality and exclude all private costs. This is a useful start that indicates the benefits of both investment and the need for further work.

The potentially large sums of NHS money that could be saved through better quality housing compare with the relatively modest amounts of investment being undertaken with government money. The new HEES has a total budget of £300m over two years, double the 1999 level of £75m.

INTER-AGENCY PARTNERSHIP IN PRACTICE

At the core of this book is the recognition that affordable warmth strategies require an interdisciplinary approach - they require a partnership between a variety of professionals. It takes time to develop the shared vocabulary and agenda, but without these it is not possible to communicate and liaise effectively. The process is slow initially, but the talking stage is an essential foundation for a more productive collaboration.

The partnerships developed in Cornwall, to help asthmatic children, are described by Ian Mackenzie and Margaret Somerville. This project has been recognised by a National Health Service "Beacon Award" for health improvement.

Local authority housing departments and housing associations were able to use some NHS money to reduce damp in the homes of children with asthma, who had been identified by their GPs. The problem of research where the children are severely ill meant that all those identified were helped, with no control group. Thus, being humanitarian and fitting in with the vagaries of funding arrangements (e.g. time pressures) limited the design of the research. In addition, local authority tenants (in poor quality housing) move frequently, so 12 families out of 98 left the houses in the first year.

Despite these constraints and concerns, 98 homes were improved at an average cost of £3061 per house. As a result there was an improvement in the number of children's bedrooms with heating (from 10% to 86%) and a decline in the number of bedrooms with dampness (from 60% to 21%). All respiratory symptoms were reduced and there was much less absence from school because of asthma. The savings to the NHS exceeded the annual equivalent of housing improvement costs, largely because of reduced hospital admissions of one individual.

Rob Howard and Roger Critchley report on a partnership with financial support from five sources, including NEA and Nottingham City Council and Health Authority. The research is being undertaken - and again it is not complete - with seven households, who were due to get full central heating on the grounds of ill health (asthma). Anecdotal results indicate that the tenants report a reduction in symptoms and in drug use (some of the drugs used to limit asthmatic attacks are expensive). The paper elucidates the benefits and pitfalls of partnership working, but is entirely supportive of the need for housing and health authorities to work together.

Helen Thompson describes the approach used in Nottingham - which includes another NHS Beacon project. A series of seminars brought people together with health, social and environmental perspectives to identify common problems. The Nottingham Health Action Group will be taking the lead on developing the 'Affordable Warmth' workstream as part of the local Health Action Zone. An important component is to train health workers to recognise signs of energy inefficient homes.

Birmingham has come up with another pioneering approach to poor quality housing, as reported by Adrian Jones. In the 1970s, the idea of 'enveloping' evolved here - making the external envelope of whole streets of houses sound and weatherproof, regardless of tenure. This new initiative also recognises the importance of a holistic approach - it is no good putting loft insulation in if the roof is leaking – and, in particular, of tackling problems at an early stage. This is the local authority acting as an enabler, encouraging self-help through community partnerships. The aim is to help residents improve their properties, before a minor fault becomes major deterioration. Services are designed to be delivered through residents' own community self-help schemes, which enables them to identify the problems they want to tackle. Access to independent advice and reliable builders have been two of the main concerns.

A wide range of assistance has been made available from outside groups: the police and security systems, fire brigade and smoke alarms, health services and safety aids. Young people from one of the schools offered a light-bulb-changing service, part funded by the local authority, because this prevented accidents amongst the elderly. By using long-lasting, low-energy bulbs, the pensioners save money and it is many years before the bulb needs replacing again. Overall, more home improvements were undertaken. In one area, there was a four-fold increase and in another a third of owner-occupiers and half of local authority tenants would not have carried out the work without the support of the Urban Care initiative. This project is not tackling cold damp homes solely or even targeting them directly, but it is part of the process of preventing them.

Emma Jones, Victoria Wiltshire and Helena Poldervaart have collated information on ten case studies, where the health and local authorities are working in partnership. The most extensive collaboration is Oxford's 'medically supported heating scheme', which has used GPs and health visitors to identify properties that need improvement.

There is still no generally accepted, robust piece of research that quantifies the health benefits of investing in energy efficiency. One problem is that health authorities are more focused on service provision than on prevention. The Health Action Zones provide an opportunity to combine resources and expertise and replicate many of these successful initiatives and learn from the pitfalls identified by the authors.

WAYS FORWARD

Now the link between poor health and energy inefficient houses has been accepted, the debate focuses on intervention time-scales and costs. There is no doubt that the problem of fuel poverty is both extensive and intensive. Even the minimum estimate, 4.4 million households in England, is extensive and the most deprived households would have to spend another £700 a year to have adequate energy services [DoE 1998, p234]. The required investment programme is assessed at well over a £1,000m pa for 15-20 years [Boardman 1991, p207]. To justify this expenditure and prompt the necessary action, it is useful to identify the costs inherent in the present situation. This does not alter the need for investment, but might increase the size of the budget. It is certainly an example of the need for the government to undertake inter-departmental analysis.

Peter Ambrose looks at the cost effectiveness of housing investment and identifies the costs falling on budgets other than housing, that would be reduced by better quality housing. These exported costs result from the diversion of specialist professional expertise, for instance doctors writing letters about housing conditions, or teachers giving social work support. It has been difficult to identify the actual costs, but he has developed the role of indices and a matrix of cost categories.

The range of the costs are identified by comparing central Stepney, where there are very strong links between poor health and poor housing, with another area of London, Paddington. The latter is an area of improved housing where the cost to the health service is only £72 p.a. per household, compared with £515 in Stepney - seven times as much. Designing out these costs requires investment in better heating and insulation along with many other recommendations. This is a necessary, but not sufficient, condition for the achievement of higher health standards.

Martin Bardsley discusses the ways in which housing and health authorities could liaise. The substantial annual costs to the NHS that result from poor quality housing are the reason why there should be collaboration. There are further important questions about whether NHS resources should be spent on better quality housing. What are the true responsibilities of the different agencies? Local housing authorities, under the Single Regeneration Budget, bid for funds and in 1998 were told that these should contribute to improvements in public health. In the previous year, only 8 out of 101 bids from London had a significant health partner. Even so, information on the health benefits of regeneration strategies is fairly limited. This is partly because some of the benefits are long-term, rather than immediately identifiable. Quantifying the benefit and calculating cost-effectiveness is fraught with difficulties. The link between poor quality housing and ill health is now accepted, but what is not known is the value in health benefits of changes in the condition of people's housing.

Health authorities are being required to be more collaborative, as well as housing authorities, with a focus on health improvement, not just on treating ill health. Housing-related issues are fragmented within the NHS, which hinders recognition of the importance of the issue. As a result, we are still some way from a common language in which health benefits can be linked to housing. Bardsley provides a matrix to aid the auditing of health authority activity in relation to housing.

In 'The Unavoidable Imperative', Professor Peter Smith gives a convincing and imaginative rallying call for the Government to take action. As he correctly states, the sums being allocated by the DETR for housing investment are well below the levels needed. Whether the future lies in energy service companies or local authority targets, it is clear that the evidence provided in this book points to just one answer: a substantial investment programme in more energy efficient homes to ensure that all have affordable warmth and healthy housing.

Excess winter mortality can be virtually eliminated if people are able to keep warm, in the home. Even exposure to cold outside is less risky if you were previously warm in the house. As a result of improvements to the energy efficiency of the home, low-income families can achieve the same standard of health as the average citizen. Previously, they were far less healthy.

For these social reasons - and parallel environmental ones - upgrading the housing stock to provide affordable warmth is an imperative.

Part One:

Current research

1
Cold weather, cold homes and winter mortality

2
Housing and winter death: epidemiological evidence

3
Cold, cold housing and respiratory illnesses

4
Cold stress, circulatory illness and the elderly

5
Dust mite allergens, indoor humidity and asthma

6
Impact of fuel poverty on health in Tower Hamlets

7
Tolerant building: the impact of energy efficiency measures on living conditions and health status

1

Cold weather, cold homes and winter mortality

W.R.Keatinge and G.C.Donaldson

The key questions addressed in this section are the nature of the relationship between cold housing and premature illness and death, and what are the most practical and cost effective ways of handling the problem. The following articles cover a number of recent studies on the effects of cold and damp housing. The purpose of this chapter is to review existing evidence about the extent and nature of effects of cold exposure, and particularly of cold housing, on health.

Awareness of seasonal mortality has a long history. In the first prose work written on the continent of Europe, in the fifth century BC, Herodotos remarked that change, more especially change of seasons, was the great cause of men falling sick. There are enough passing references in the literature since then to show that a link between cold weather and deaths, particularly from pneumonia, was generally accepted throughout the 24 centuries since that time. Statistical evidence of the extent of the problem has been available for more than a century and a half. The most secure measure of serious effects on health is provided by changes in daily deaths, and British mortality records since 1841 have shown these to rise strikingly in winter.

However, it was not until the 1960s and early 1970s that there was any widespread recognition of these deaths as a major public health problem that called for serious scientific investigation and effective preventive measures. Although there were reports throughout that time of deaths from coronary and cerebral thrombosis, as well as pneumonia, increasing in cold weather [see Bull and Morton, 1978], there was a widespread assumption in Britain that excess deaths in cold weather must be due to hypothermia, simple cooling of the body core. It was supposed that indications to the contrary could be explained by failure to recognise hypothermia.

That suggestion was plausible, since ordinary clinical thermometers generally failed to read below 35°C, the official boundary point for hypothermia. Use of low reading thermometers produced substantial numbers of readings below 35°C in apparently healthy people in their homes. However, these turned out often to be misleading readings from the mouth, resulting from the mouth filling with cold saliva when the face was cold [Keatinge and Sloan, 1975]. Use of urine temperature can also give false low readings if the urine stream is poor. A study was therefore made in which mouth temperature was measured throughout a year in all emergency patients admitted from their homes to the London Hospital and associated hospitals in the eastern part of London. Any readings below 36°C were checked by a rectal reading. The study showed only 14 people with body core temperatures below 35°C. All of these had some other serious illness, and had usually been incapacitated by this illness, and became hypothermic as a result of that [Woodhouse et al., 1989]. Hypothermia is in fact rare even among people living in poorly heated homes, and the evidence of death certificates that shows arterial thrombosis and respiratory disease as causing most of the excess winter mortality can be taken as reliable.

A key question is whether the excess mortality in winter is due to direct effects of cold on people, or to indirect factors such as lack of vitamin C in the winter diet. Close short term associations between mean daily temperature and mortality have provided strong evidence for direct effects. These have been shown in various ways, but the simplest is to use simple regressions to calculate the average pattern of daily temperatures associated with a cold day for the time of year in winter, and then the patterns of daily mortality associated with this [Donaldson and Keatinge, 1997]. Figure 1.1 shows the results, with all cause mortality rising as temperature falls, and then falling a few days after temperature starts to return to normal. Thrombotic and respiratory deaths also rose, but with different time courses. The delay of peak deaths on peak cold was no more than three days in case of deaths from coronary thrombosis, but twelve days in the case of respiratory disease. Reduced vitamin C intake and other indirect factors probably also contribute to winter mortality [Khaw K-T and Woodhouse, 1995], but rapid and relatively direct effects of cold seem to play the leading part.

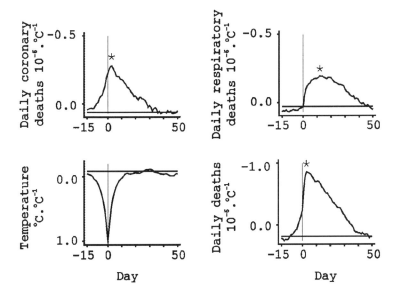

Fig. 1.1. Time courses of cold spells and associated mortality in south east England. Deseasonalised data in linear range 0-15°C   * p<.001. (J Epidem Comm Health 1997; **51**: 643-8. Figure 3 from Donaldson and Keatinge).

The nature of these effects is reasonably clear in the case of thrombotic disease. A normal part of the body's defence against cold is to shut down blood vessels in the skin, reducing heat loss from the body core. However, this displaces around a litre of blood from the skin, and this overloads the circulation in central organs of the body. In order to reduce this excess volume of blood, salt and water is excreted, and further salt and water leaves the circulation through the walls of the blood capillaries to enter the intestitial spaces. This adjusts the blood volume to the reduced capacity of the circulation, but leaves the blood more concentrated [Keatinge *et al.* 1984, Neild *et al.* 1994]. Some of the smaller components of the blood plasma, including the anti-thrombotic protein C, are able to redistribute through the capillary walls but the red blood cells, platelets, white cells, and fibrinogen are too large and remain in increased concentration in the plasma (Figure 1.2). All of them promote clotting. This can be detected among the general public in cold weather [Donaldson *et al.* 1997] as well as during experimental exposures of volunteers to mild cold. It does little harm to anyone with good condition, but people over middle age have increasing degrees of atheroma, and the concentration of the blood in cold weather greatly increases the risk of a clot forming in one of their arteries. It is therefore the older members of

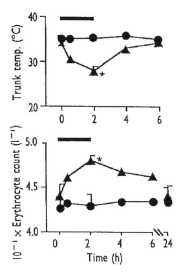

Fig. 1.2. Increased red cell count caused by cold exposure in elderly volunteers.
▲ Exposure to cold turbulent air at 18°C ● Controls   n =6   * p<.05
Reproduced with permission from Neild *et al.*, 1994, Clinical Science; **86**: 43-48; the Biochemical Society and the Medical Research Society).

the population who are most at risk of strokes and heart attacks in cold spells.

Another reason for the increase in thrombosis is that respiratory infections in winter induce an acute phase response, with rise in plasma fibrinogen which also promotes arterial thrombosis [Woodhouse *et al.*, 1994]. The reasons for the increase of respiratory infections and respiratory deaths in cold weather are not fully understood, but they probably include bronchoconstriction induced reflexly by cooling of the face and airways [Millqvist *et al.*, 1987, Giesbrecht 1995], local cooling of the respiratory tract, and suppression of immune responses to infection by the adrenal cortex. (Cold and respiratory illnesses are discussed in detail in Chapter 3.) A few sudden deaths from coronary artery disease are probably also caused by the increase in arterial pressure that cold exposure induces, as referred to in Chapter 4.

In order to assess the most effective means to reduce winter mortality, it is important to know whether the cold exposure causing it is experienced mainly indoors or outdoors. Evidence that outdoor cold can be important was provided by a study of people living in fully heated housing which was managed by Anchor Housing. This showed that despite the homes being heated to full comfort level, the residents, who spent considerable time outside, had a substantial increase in mortality in winter [Keatinge, 1986]. Study of representative sections of the elderly population in a wide range of European countries gave an opportunity to evaluate the relative contributions of the two, and indicated an approximately equal contribution of cold housing and outdoor cold to excess winter deaths [Eurowinter, 1997].

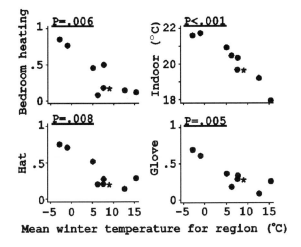

Mean winter temperature for region (°C)

Fig. 1.3. Home heating and clothing in cold and warm parts of Europe, at standard outdoor temperature of 7°C . Points from left to right: north Finland (NF), South Finland (SF),Baden-Wurttemburg (BW), Netherlands (N), London (L), north Italy (I), Athens (A), Palermo (P). London given asterisk to distinguish it from north Italy. (From Keatinge, Donaldson, Bucher, Jendritsky, Cordioli, Martinelli, Dardanoni, Katsouyanni, Kunst, Mackenbach, McDonald, Nayha, Vuori; Lancet 1997; **349**: 1341-6).

The Eurowinter study also gave an opportunity to assess how effectively people protect themselves from winter cold and from winter mortality in the widely different climates within Europe. To do this, it included direct measurements of home temperatures, and a questionnaire survey of time spent outdoors, and clothing worn and physical activity during outdoor excursions. These covered 1000 people in each of eight regions from Athens in the south to Finland in the north. The results showed that people in cold climates, not surprisingly, protected themselves from cold much more effectively than those in warm climates. What was surprising was that, even at a given level of outdoor cold, the people in cold climates had warmer houses, wore warmer clothes, and when outdoors kept moving, and were much less likely to feel cold enough to shiver (Figure 1.3). Throughout the winter there were in fact fewer excess deaths per million people in the selected age groups (50-59 and 65-

74 years) in Finland than in London or Athens despite much lower winter temperatures in Finland. The main differences outdoors were much more frequent wearing of hats and gloves in the colder countries, and striking differences indoors were higher living room temperatures and more frequent bedroom heating in the colder countries, all at a given level of outdoor cold.

Such adjustment to cold conditions was seen to an extreme degree in people living in Siberian cities, where temperatures were lower than in any part of western Europe. A survey in these similar to the Eurowinter study showed that as temperature in Ekaterinburg, a large industrial city north-east of Moscow, fell to 0°C, people increased their level of clothing and physical activity outdoors, and their bedroom heating indoors; reports of shivering, and mortalities, did not increase [Donaldson *et al.*, 1998c]. Only when outdoor temperature fell below 0°C did the defensive measures reach a plateau, and mortality then rose as temperature fell to -29.6°C. Yakutsk, in eastern Siberia, has the lowest winter temperatures of any city in the world, and showed the most dramatic example of adjustment of a population to cold [Donaldson *et al.* 1998b]. The most striking of these was wearing of very thick outdoor clothing, mainly of fur. Yakutsk was the only region surveyed in which people reduced their number of outdoor excursions when the temperature fell. They only did so when it fell below -20°C, to -48.2°C. Despite the extreme cold, there was no increase in all-cause mortality in Yakutsk in winter. A small rise in respiratory mortality as temperature fell was balanced by a reduction in deaths from injury.

The fact that adequate protection against cold can virtually eliminate excess winter mortality in the coldest parts of the world presents the obvious challenge that it should be possible to eliminate it elsewhere. It may be that the populations of cold regions owe some of their immunity to inborn or acquired adaptation, but the close temporal associations elsewhere between cold exposure and mortality, and the much more effective defensive measures taken against cold in the colder regions, point strongly to these as the main factor protecting against winter deaths. The recent breakdown of mortalities and other data in Britain by wards has given an opportunity to seek associations of social factors in different regions of the country with winter mortality. This offers possibilities of new perspectives and is described in Chapter 2, although it is difficult to establish differences between wards within a small country where conditions are much more uniform than across continents.

The fact that clear statements can now be made to the general public, on the need for effective protection from cold, may be one of the most important results of the recent studies. An obvious practical problem is that, while no-one in Yakutsk needs to be persuaded to give high priority to keeping warm in winter, the need for it is much less obvious to people in London or Athens. Elderly people, who are most at risk, do not always take kindly to what can sound to them as obvious or patronising advice. They are much more responsive if told that most people in their area do not take that advice, and have a high death rate as a result.

There is also, of course, a major role for voluntary and governmental action in reducing winter mortality. Continued action to promote energy efficient homes, which keep people warm without adding to global warming and pollution by burning unnecessary fuel, remains important. It is necessary to recognise some dilemma here, as crowded homes are energy efficient in terms of fuel expenditure per person, but excessive crowding causes other problems. The need for a balance in home occupancy levels, and perhaps for homes to have core areas with fully effective heating, and peripheral areas that lack heating but can be used in warm weather, presents challenges for the architect. The need to keep humidity low and so reduce the risk of asthma due to house mites when homes have been well insulated, is another problem for architects, and is addressed by Dr Howieson in Chapter 5.

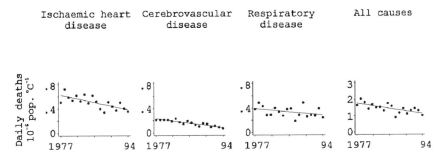

Fig. 1.4. Decline in excess winter deaths. All of the causes of excess winter deaths in Greater London show declines, p < .01. (This figure was first published in the BMJ 1997; **315**: 1055-6, Donaldson and Keatinge, and is reproduced by permission of the BMJ).

Recent progress in reducing winter mortality in Britain has been real. Substantial falls in winter mortality up to 1977 were partly due to disappearance of massive winter epidemics of influenza that caused large numbers of such deaths up to that time [Keatinge *et al.*, 1989]. Much of that may have simply resulted from less influx of new and lethal strains of the virus. Since then improvement in winter mortality has been slower. Analysis is complicated by the increase in life expectancy, with increase in numbers of the elderly people who are most vulnerable to winter mortality. However, if allowance is made for this by analysing specific age groups (Figure 1.4), it is clear that people's risk of being added to the winter toll has decreased steadily over the last 20 years [Donaldson and Keatinge, 1997]. Credit for this should probably be divided between organised measures to improve warmth in the home and to advise on personal protection against cold by clothing and physical activity when outdoors, and general increase in prosperity that makes it easier for people to avoid unwanted personal exposure to cold. Nevertheless, there are still around 40,000 excess winter deaths in Britain every year, the highest rate in Europe. These deaths are essentially preventable, and much remains to be done.

2

Housing and winter death: epidemiological evidence

Paul Wilkinson, Megan Landon and Simon Stevenson

2.1 Introduction

In Britain, close to 40,000 more deaths occur between December and March than expected from the death rates in other months of the year [Curwen, 1990]. Although deaths from many causes show a winter excess, the principal increase is in cardiovascular and respiratory mortality [Curwen, 1990; West, 1989; Enquselassie et al., 1993; Knox, 1981; Fleming et al., 1993]. A similar pattern is seen in other European countries [Crombie et al., 1995; Saez et al., 1995; Kunst et al., 1993] but, at around 20%, the magnitude of the winter excess in Britain is comparatively large [Curwen, 1990; McKee, 1989].

As discussed in Chapter 1, the factors that may contribute to the excess are numerous. Climatic shift is an obvious and important element, but winter is also associated with more respiratory infections, with behavioural changes, with changes in the pattern of indoor and outdoor air pollution, and with small, but perhaps significant, shifts in the intake of micro-nutrients such as vitamins C and D. These micro-nutrient changes may influence the risk of acute cardiovascular events in particular [Khaw et al., 1995; Ness et al., 1996]. Among the seasonal infections, influenza makes one of the largest contributions. Even in non-epidemic years, it probably accounts for 3,000 to 4,000 winter deaths nationally, and for many more in epidemic years, such as during the winter of 1989/90 [Ashley et al., 1991].

Though the relationships of these factors with excess winter mortality are not precisely quantified, epidemiological evidence suggests that low ambient temperature accounts for the greater part of the seasonal increase in deaths [Eurowinter Group, 1997; Wilmshurst, 1994; Khaw, 1995; Curwen, 1990]. Moreover, cold exposure is known to affect a range of haemodynamic and haemostatic parameters [Keatinge *et al.*, 1984; Neild *et al.*, 1994; Woodhouse *et al.*, 1993; Stout *et al.*, 1994], and may also influence susceptibility to infection.

Given the primacy of cold exposure in excess winter death, it is incumbent upon those concerned with public health to ask how its health impacts can be ameliorated. One aspect of this question is the role of housing, and whether improvements to house conditions would reduce the number of winter deaths. Though as yet there is little direct evidence that housing has an appreciable influence, extrapolation from what is known about patho-physiological mechanisms would suggest that cold homes are likely to have an appreciable impact on winter mortality, at least in the most vulnerable groups.

In this chapter we examine some of the epidemiological evidence relating to the seasonal fluctuation in mortality and its potential relationship with house conditions and home heating.

2.2 Seasonal variation in mortality

It is not surprising that deaths are more common in winter months than at other times of the year. It is to be expected, for example, that some vulnerable individuals with advanced cardio-respiratory illness may succumb to the additional insult of respiratory infection or cold, and such deaths are bound to predominate in winter months. What to many is surprising, is the magnitude of the seasonal fluctuation in mortality in Britain.

Figure 2.1 shows daily counts of deaths in London for the period 1986 to 1996. There is variation in deaths from day to day, much of which can be ascribed to random fluctuation, and some variation in the pattern of deaths from year to year. The epidemic of influenza in the winter of 1989/90 accounts for the short but particularly large rise in daily deaths in this winter. Daily deaths at the peak of this epidemic were more than double the non-winter baseline.

But the most conspicuous feature is the marked oscillation in the underlying mortality risk across the year. The continuous line represents the two-month moving average of deaths, which thus smoothes out the sharpest changes in mortality, blunting any short-lived peaks and troughs. It is therefore remarkable that, over this eleven-year period as a whole, the mean winter peak of this curve is more than 40% higher than the mean of the summer troughs. In other words, for every ten deaths which occur at the lowest rate of deaths in summer (taken over a two month period), on average more than 14 occur at the maximum of the winter death rate.

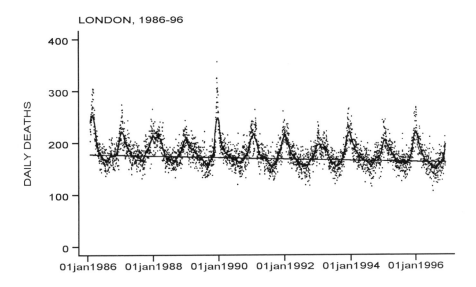

Fig. 2.1. Daily count, two month moving average ▬ and non-winter baseline ⎯ of deaths in London 1986-1996.

This pattern also shows that the conventional definition of excess winter death — the percentage by which the mortality rate for the period December to March exceeds that of other months of the year — does not reflect the true magnitude of variation in death rates across the year: 20% is simply the figure which derives from applying an arbitrary definition of winter based on calendar months. Even so, a 20% excess is a very large risk applied to the population as a whole. There are few specific risk factors that have such an impact on death rates at population level and in numerical terms the seasonal increase in deaths of cardiovascular and respiratory deaths, is substantially greater than that attributable to many individual health hazards.

A further point to note is that the underlying mortality rate fluctuates throughout the year, rather than a step increase in the winter period. Plots of mortality against maximum daytime temperature show that the daily mortality rate increases more-or-less linearly for each degree fall in temperature below 20 degrees Celsius. Thus, it is not just the very coldest days that are associated with higher death rates; an increase above the minimum rate is apparent even at quite moderate ambient temperatures. This suggests either that humans are very sensitive to temperature fluctuations or that there are other seasonally varying factors that account for the seasonal fluctuation in deaths. At the highest ambient temperatures (daily maximum above 30 degrees Celsius) death rates increase, mainly attributable to the effects on the cardio-vascular system of heat stress.

2.3 Relationship with housing

The relative size of the seasonal swing in mortality in Britain has drawn comment from various observers. McKee was among the first to point out the paradox that Britain's winter excess is among the largest in Europe despite the fact that Britain is buffered against extreme cold by the Gulf Stream and surrounding seas [McKee, 1989]. Given the comparative mildness of our winters one would predict any temperature-related impact on health to be correspondingly modest. In fact, the reverse appears to be the case. The starkest contrast is with the Scandinavian countries which have small increases in winter death though their winters are very much harsher than in Britain [McKee, 1989; Laake and Sverre, 1996]. There may be several reasons for this paradox, and it might in part be an artefact of the way the winter period is defined: the seasonal fall in ambient temperature occurs very much earlier in Scandinavia than in Britain, and the conventional comparison of December to March as winter against other months may therefore be based on different parts of the climatic cycle in different countries. Nonetheless, the relative sizes of the winter excess in Britain and Scandinavia are unexpected, and the observation has led to speculation about such factors as the importance of the dampness of our climate, about behavioural factors — the failure to take sensible precautions to protect ourselves against the cold — and about the various factors that contribute to individual vulnerability.

     As Brenda Boardman argued in her book 'Fuel Poverty' [Boardman, 1991], one of those vulnerability factors may be housing quality. Compared with many north European countries British housing has low thermal efficiency. Certainly, homes in Britain are not built to the same standards of thermal protection as those in Scandinavia, chiefly because there is less need to do so. The consequence may be that indoor temperatures are less well maintained, leading to exposure to sub-optimal ambient temperatures. People above retirement age in particular spend the great majority of their time at home, and if that environment is cool or frankly cold, it is quite possible that it will result in peripheral and core temperature chilling with adverse consequences for health.

The policy relevance of quantifying the extent to which poor housing contributes to vulnerability to cold is obvious. Unfortunately, as yet we have little direct evidence that enables this to be done, though we may conclude from indirect evidence that housing is likely to be important. To obtain estimates of excess winter death associated with poor housing we need to combine detailed information on characteristics of a very large number of dwellings, including information on their thermal efficiency and space heating, with the mortality statistics of their inhabitants. The epidemiological question is the variation in the seasonal increase in mortality across different housing conditions. It is hoped that current research using quantitative risk assessment methods will provide such estimates. These estimates will inevitably entail scientific judgement about the degree to which observed statistical associations between housing quality and excess winter death represent cause and effect. Direct evidence about the health benefits of energy-related housing interventions is also needed, and will require large scale before-and-after studies.

An indication of the potential importance of house conditions comes from analyses of mortality in census wards of South East England which found that the proportion of homes without central heating was one of the strongest predictors of geographical variation in excess winter death [Wilkinson et al., 1998]. The ward-level Carstairs deprivation score [Carstairs and Morris, 1991] was found to be a comparatively weak predictor, the winter excess of deaths in affluent wards being almost as great as that in the most deprived wards — a surprising finding given the strong deprivation gradient seen in absolute deaths rates for many specific causes and for all-cause mortality as a whole. In these analyses ward markers of the proportion of local authority housing, the proportion of elderly living alone, and population density individually were weakly associated with excess winter death, but not when adjusted for age, deprivation and proportion of homes without central heating.

The results suggests that around one per cent of all deaths in winter are attributable to lack of central heating. A ward marker of central heating is clearly a very crude measure of the adequacy of home heating, but the fact that it is associated with winter death is plausible, and it suggests that significant impact may be obtained from measures to improve the thermal efficiency of houses and the affordability of heating them.

2.4 Affordable warmth

Although some uncertainties regarding the relationship between housing and winter death remain, the potential health hazard of low indoor temperatures is recognized and is discussed further in Chapter 3. A comfortable indoor temperature is around 21°C. Below 18°C there begins to be some discomfort, and the risk of adverse effects – respiratory infections, bronchitis, heart attacks, stroke – rises. The risk increases the more temperature falls. Below 10°C the risk of hypothermia becomes appreciable, especially for the elderly.

It is therefore a reasonable goal for housing policy to ensure that all households, especially those containing vulnerable individuals, are able to heat their living space to around 21°C, and are able to do so without compromising other important areas of household expenditure. The standard heating regime specified by the Department of the Environment Transport and the Regions is 21°C for the living room and 18°C in other rooms, and a minimum heating regime specified as 18°C in the living room and 16°C in other rooms.

Much useful data on heating patterns are provided by the Energy Report of the 1991 English House Condition Survey (EHCS) [DoE, 1996]. In 1991/92, though the winter was mild, only 25% of homes had measured internal temperatures which met those of the standard heating regime, and only 70% met the temperatures of the minimum regime. On days when the outdoor temperature was below 4°C, the proportion of homes failing to meet even the minimum regime rose to 50% in owner occupied dwellings, 62% in council homes and 95% in the private rented sector.

As these figures suggest, not only is there an apparent deficiency in home heating but also an inequality in it. Among the factors that govern the adequacy of home heating (Figure 2.2), the most important are the size and thermal efficiency of the dwelling, the effectiveness and use of the heating system, and household income. With all three, it is the most vulnerable who are often the most disadvantaged.

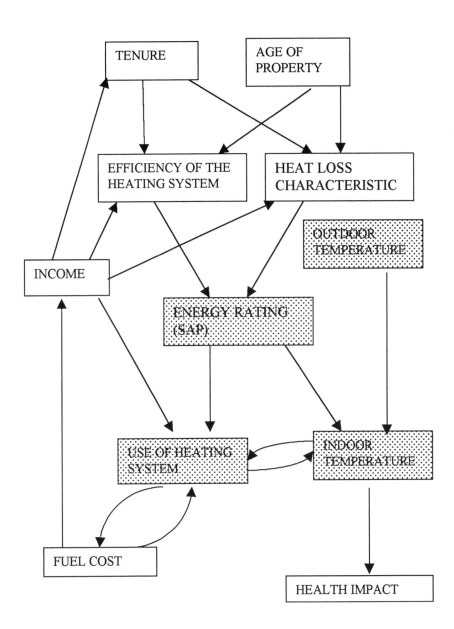

Fig. 2.2. Schema of the inter-relationship between poverty, energy efficiency, heating costs and indoor temperature

The usual definition of the energy efficiency is the Standard Assessment Procedure (SAP) [DoE, 1994], which reflects the cost of heating the dwelling to the standard regime. It is calculated on a logarithmic scale of 1 (highly inefficient) to 100 (highly efficient) and is normalized for floor area. It is a function of the heat loss/heat gain characteristics of the building, and the efficiency of the heating system. The greater variation is in the heat loss/heat gain characteristics, the principal determinant of which is building age as this governs the standards of construction.

Table 2.1. Number of households (thousands) in fuel poverty, 1991 and 1996

		Percent of household income needed to be spent on fuel		
Year	*No. of households*	*<10%*	*10 - 19.9%*	*>20%*
1991	19,111	12,482 *65.3%*	4,360 *22.8%*	2,270 *11.9%*
1996 estimates (same basis as 1991 EHCS)	19,643	14,367 *73.2%*	4,092 *20.8%*	1,184 *6.0%*
1996 estimates (including housing costs in calculation of household income)	19,643	15,271 *77.8%*	3,598 *18.3%*	774 *3.9%*

Source. Adapted from: *Fuel poverty: the new HEES — a programme for warmer, healthier homes. DETR, 1999. p11.* [Crown copyright: reproduced with the permission of the Controller of Her Majesty's Stationery Office]

Estimates derived from the 1991 EHCS survey are that three million dwellings have SAP ratings below 20. But there is a gradation with owner occupiers having appreciably better ratings than those in the social rented sector, who in turn have better average ratings than private tenants. Among the latter, one of the groups with least energy efficient dwellings is lone pensioners — a key risk group for excess winter death — who have an average SAP rating of just 11. Even in other housing sectors, pensioners tend to live in dwellings of poor energy efficiency, such as bungalows and older terraced properties. A quarter of those aged 75 years or more live in dwellings with SAP ratings below 20.

Clearly, dwellings that are more energy efficient are cheaper to heat and can more easily be maintained at desired temperature, even with only intermittent heating. However, fuel consumption is generally lowest in the least energy efficient homes, in part because they tend to be smaller, but also because they are more often occupied by pensioners and households of low income. Such households tend to have more expensive heating systems, such as those which use on-peak electricity, and their homes tend to be less adequately heated. The cost of home heating is evidently a major determinant of fuel use, and this is one of the principal reasons that fuel consumption is generally greater in dwellings of higher (more energy efficient) SAP ratings as on the whole these are occupied by households of higher income.

The concept of affordable warmth has been defined by Boardman on the basis of the proportion of household income which would need to be spent on energy services. The common definition of fuel poverty is when a household would need to spend more than 10% of its income on fuel as a whole when such fuel expenditure includes heating the home to the standard regime [Boardman, 1991]. By this definition around one in three households in the 1991 EHCS could be categorized as in fuel poverty, a figure which has fallen to around one in four by 1996 (Table 2.1).

From a public health perspective, benefits could be expected if the proportion of households in fuel poverty could be reduced through measures to improve the energy efficiency of both current and new housing stock, or by measures to help with fuel payments (Winter Fuel Payments, Cold Weather Payments, reduction in fuel taxes). The decrease in fuel poverty between the 1991 and 1996 English House Condition Surveys is encouraging, but the fact that even in the later survey 4.4 million households were estimated to need to spend more than 10% of their income on fuel indicates the scale of the problem.

As a remedy there are advantages to concentrating effort on improvement in the energy efficiency. Not only is this a more lasting solution for improved homes, but it is easier to control the indoor temperature once such improvements have been made, hence making it more likely that satisfactory indoor temperatures can be achieved, and energy consumption is diminished, with benefit to the environment. In many cases it is estimated that the cost of implementing energy efficiency improvements could be recovered within a few years from reduced fuel costs, though in practice households tend to spend part of the efficiency gain on having a warmer home rather than on lower fuel bills [Milne and Boardman, 1997].

The UK Government's proposed new Home Energy Efficiency Scheme (HEES) [DETR, 1999] specifically targets energy efficiency improvement on the most vulnerable groups in private rented accommodation and owner occupiers. The rented sector includes the highest *proportion* of fuel poor households, and the owner occupied sector the largest *number* of fuel poor households. The main thrust in relation to social housing is through an increase in the capital resources available to local authorities for housing. There is also a role for assistance with fuel bills for households in greatest need, and some steps have been taken in this area, with the increase in winter fuel payments for pensioner households from 1999/2000.

2.5 Impact on excess winter death of housing improvements

The benefits that accrue from home energy improvement measures are difficult to determine. The HEES initiative has proved popular and those who have had insulation work carried out under the scheme report significant gain in comfort . There is little doubt of their impact on quality of life. However, for reasons alluded to above, the impact of such schemes on winter death is more difficult to ascertain.

It is also worth noting that winter mortality is not just a problem of fuel poverty. In small area analyses, only very modest variation is found in the magnitude of excess winter death between rich and poor areas [Wilkinson et al., 1998]. This suggests that excess winter death is far from being a problem confined to fuel poor households. Though such households may indeed be particularly vulnerable, perhaps the greatest number of excess winter deaths occur in households without fuel poverty (which are more numerous).

Keatinge and colleagues have pointed out that, over past decades, there has been a substantial increase in use of central heating and indoor temperatures, yet little decline in winter excess deaths from cardio-vascular disease [Keatinge et al., 1989]. These authors suggest that such deaths are more likely to be due to the effects of brief outdoor excursions into the cold than to a change in average indoor temperature. It should also be remembered that only a proportion of winter deaths is attributable to low temperatures, and that proportion remains imprecisely defined. Nonetheless, a link between indoor temperature and winter death remains probable given our wider knowledge of the patho-physiology of cold-induced illness (Chapter 3; Chapter 4), and there is little doubt that home energy improvement measures can lead to gains in indoor temperature that might well translate into significant health benefits. How great those benefits are remains to be seen.

2.6 Conclusions

Britain has a large winter excess of mortality, especially in the elderly age-group. Much of the seasonal fluctuation is attributable to the effects of cold, though the precise proportion remains uncertain. Paradoxically, given our mild climate, Britain has an apparently greater winter excess than many other European countries for reasons that remain unclear, though poor housing may well be one factor.

This possibility is highlighted by the fact that around some four million households in England live in fuel poverty – that is, they would have to spend more than 10% of their income to achieve a standard heating regime yielding temperatures of 21°C in the living room and 18°C in other rooms. Such households are concentrated in disadvantaged groups, especially people of retirement age living alone. These are one of the groups at greatest risk of excess winter death.

Though conclusive proof is not yet available that improvement in housing conditions would reduce winter mortality, indirect evidence is persuasive, and steps to raise the energy efficiency of new and existing stock can be expected to have an impact. The primary research need is to assess the short, medium and longer term health effect of such housing interventions, and how they can best be targeted to greatest effect. Some of this need is beginning to be addressed, as later chapters demonstrate.

3

Cold, cold housing and respiratory illnesses

Ken Collins

3.1 Introduction

From the earliest Registrar General's statistics and epidemiological studies in the UK an association has been established between cold ambient temperatures and respiratory illnesses. The invariable winter rise in the incidence of respiratory diseases such as bronchitis and broncho-pneumonia is ascribed to the adverse effects of cold magnified by atmospheric pollution [Wright and Wright, 1945]. Cold conditions *per se* are not likely to result in respiratory infections in the absence of respiratory pathogens, as for example shown by the study on the island of Spitzbergen where throughout a severely cold winter no common colds or respiratory infections occurred until the arrival of the first ship at the end of the winter [Tyrrell, 1965]. However, a marked reduction in the temperature of the respiratory tract can increase susceptibility to pathogens by affecting the muco-ciliary defences and by initiating local inflammation.

In the present review the main consideration is given to cold indoor conditions. Moderately cold indoor temperatures where dampness is present poses a common health hazard engendered by the growth of mould and fungi capable of causing allergy and respiratory infections. The completion of several epidemiological studies recently may now help to shed some light on the respective contributions of indoor and outdoor cold to seasonal respiratory and cardiovascular disease and mortality. Associations between the levels of environmental cold exposure and illnesses in populations may offer spurious explanations due to chance, bias and confounding. Apparent causal relationships are often prone to

confounding by social, cultural and behavioural factors and are difficult
to prove. On the basis of more recent laboratory and epidemiological
studies this review will consider the role of ambient temperatures in cold-
related respiratory illnesses, cold housing and respiratory health, and
respiratory disease as a component of excess winter mortality.

3.2 Cold environment and the respiratory system

Cold temperatures produce measurable physiological changes in the
respiratory tract through cooling and drying of the mucosal surfaces. At
rest, with ventilation through the nose, inhaled air is heated and fully
saturated before it enters the lower airways. However, when ventilation
rate and volume increase as in exercise, the breathing pattern tends to
change from nasal to oro-nasal with the result that the cooling and drying
stimulus becomes greater and moves toward more central regions of the
respiratory tract. Many recent investigations on breathing cold air deal
with the adverse responses of the lower respiratory system to cold, dry air
inhalation [Latvala et al., 1995] although the upper airways region is the
most common area affected by respiratory inflammation and infections
associated with cold.

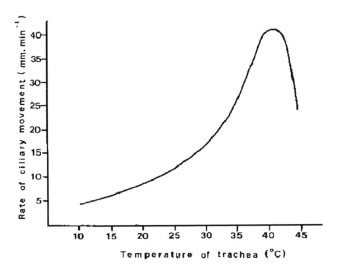

Fig. 3.1. The effect of temperature on the rate of transference of dust particles
on the ciliated epithelial surface of a strip of excised trachea. Source: Hill,
1928.

Leonard Hill [1928] classically demonstrated in mammals how the vigour of ciliary movement conveying particles over the tracheal surface was greatly decreased by local cold temperature (Fig 3.1). Other physiological effects include a cold-related decrease in the magnitude of resting membrane potential in airway smooth muscle [Souhrada and Souhrada, 1981] and a reduction in the rate of tension development in tracheal smooth muscle [Stephens, Cardinal and Simmons, 1977]. Pulmonary mechanisms are compromised by bronchoconstriction, airway congestion, increased mucous production, decreased muco-ciliary clearance and a decrease in airways mucosal blood flow [Giesbrecht, 1995]. Cold exposure is linked with impaired lung function as measured by forced expiratory volume, an effect shown to be independent of smoking habits [Rasmussen *et al.*, 1991].

Cold is an important trigger of bronchoconstriction in patients with asthma or chronic obstructive airways disease (COPD). Bronchoconstriction may occur through two basic mechanisms, firstly by the direct effect of cold inhaled air on the lower airways when the heat exchanging capacity of the upper airways is overcome, and secondly by a neural reflex mechanism initiated by facial cooling [Koselka and Tukiainen, 1995]. In cold-induced asthma, the direct effect of cold air on the airways is the most potent factor. Severely cold temperatures can initiate a lower airways inflammatory reaction. Broncho-alveolar lavage performed on healthy non-smoking subjects after 2 h exposure to -23° C showed increasing numbers of granulocytes and macrophages in the lavage fluid [Larsson *et al.*, 1998].

Although the incidence of the common cold increases in winter, cold winter temperature does not in itself appear to be the cause. The view that resistance is lowered by "chilling" in cold conditions was not supported by early experimental evidence [Tyrrell, 1965]. There have been several American volunteer studies in which core temperature was reduced but susceptibility to known rhinovirus exposure was not detectably changed. Determinants of yearly epidemics of the common cold have not been established, but certainly they include human behaviour, with more virus transmitted by higher indoor contact in winter months.

3.3 Cold housing and respiratory health

It is well established that fluctuations from an "ideal" range of hygrothermal conditions within dwellings and work places pose a threat to health, especially for certain vulnerable groups such as the elderly, infants, the sick and the disabled, all of whom spend much if not all of their time indoors [Collins, 1993; Raw and Hamilton, 1995]. Indoor hygrothermal conditions are themselves influenced by physical factors such as external climate, building structure, control systems and appliances inside the building, and by the occupants. The evidence relating to health problems comes from four principal sources:

(1) health and housing conditions (variations in health between areas with different housing conditions or over time as housing conditions change)

(2) seasonality (variations in morbidity and mortality between cold and warm seasons, changes in seasonal variations over a period of years, and differences in seasonal variations in different countries)

(3) health and weather (correlation between outdoor temperature and reported incidence of specific medical conditions)

(4) physiological investigations (studies of the effects of controlled ambient temperature on physiological function).

Though not all epidemiological studies show simple relationships between cold, damp housing and poor health, it is widely accepted that living in such housing is unhealthy and may be associated with respiratory diseases. Most studies have focused on damp and mouldy living conditions rather than cold house temperatures.

3.4 House temperatures in Great Britain

Present thermal environment standards in use in Great Britain and other countries are largely for the work- place and therefore tend to be based on the minimum conditions that should be attained and maintained for the health and well-being of groups who may be at high risk. The adverse effects of indoor cold temperatures on respiratory health studied in the work-place suggests that a critical temperature is 16° C, below which it is believed that work involving standing still should not be continued because of the increased risk of respiratory illness [Offices, Shops and Railway Premises Act, 1963]. Though this empirical level is much quoted, the data on which the statement is based are not available. There have been many large scale surveys of temperatures in homes in Britain, with spot surveys showing many houses with the warmest room below 16° C. The most recent English House Condition Survey shows that there continues to be an upward shift in temperature of dwellings in keeping with greater use of central heating. There are, however, many dwellings where the warmest room is below 16° C (2.5 million households) and even below 9°C (118,000) when the outside temperature is below 5° C [DETR, 1996] (Table 3.1).

Table 3.1. Distribution of households by temperature bands when external temperature is below 5° C, for 1986 and 1996

Temperature (° C)	Warmest room 1986	1996	Coldest room 1986	1996
24.0 plus	1.6%	2.3%	1.3%	0.7%
21.0 - 23.9	12.1%	21.0%	8.6%	10.2%
18.0 - 20.9	31.5%	42.8%	24.6%	31.0%
16.0 - 17.9	27.7%	21.4%	22.1%	25.6%
12.0 - 15.9	22.3%	10.7%	28.2%	24.9%
9.0 - 11.9	4.2%	1.2%	9.4%	4.2%
Under 9.0	0.7%	0.6%	5.9%	3.4%
Total households (thousands)	18063	19643	18063	19643

Source: English House Condition Survey, 1986 and 1996.

3.5 Hygrothermal conditions, mould growth and allergies

In cold winter conditions the moisture content of outside air (absolute humidity) is low because of moisture condensation even though the relative humidity may be 100%. When this air infiltrates the indoor environment and becomes warm it will usually create a much lower relative humidity, not much below about 35% in the UK. The warmer the building space the lower the relative humidity. Conversely, in cold houses in wet, moderately cold outdoor conditions, the indoor air may contain more moisture which may be further cooled by cold indoor surfaces upon which moisture condenses and becomes a focus for the formation of moulds. Homes with condensation problems in the UK are on average 0.5° C cooler mostly due to a 0.7° C difference in upstairs temperature [Collins, 1993]. It is usually accepted that a relative humidity of 70% is sufficient to sustain mould growth.

A quite different situation is found, however, in those regions of the world such as N. America and N. Europe with very cold climates. Indoor heating is generally very efficient, often resulting in extremely low relative humidity, and this dry environment has irritant effects on respiratory mucosal surfaces. The annual peak of respiratory illnesses in winter has been partly attributed to the decrease in indoor relative humidity when house temperatures are kept high [Green, 1982]. Evidence from several studies in humidified and non-humidified buildings tend to support the view that the occurrence of upper respiratory tract infections increases when indoor relative humidity is below 30%.

The growth of moulds in buildings depends primarily upon the humidity of the indoor environment. It is the relative humidity of the surfaces (substrates) which is of greatest significance but the relative humidity of the air directly influences the water activity of the substrate. The 1991 English House Condition Survey showed that 22% of householders reported mould growth in their homes (15% in the 1996 Survey). Mould growth is capable of causing respiratory illness through Type I and Type III allergy as well as respiratory infections. Type I allergy underlies the rapid development of symptoms such as asthma following inhalation of allergens to which the individual becomes sensitized. On present evidence, only Type I allergy can be considered to be a potential problem in a significant number of buildings [Raw and Hamilton, 1995]. In Bengali people in E. London, measured and reported low temperatures in the home were closely associated with symptoms of "hidden asthma" and that people in cold, damp houses were twice as likely to have poor respiratory health [Hyndman, 1990]. Although mould growth is widespread in homes, the evidence at present is against mould allergy being the most important mechanism linking damp housing with asthma [Strachan, 1993]. The independent contribution of mould exposure to asthma is very difficult to quantify but is likely to be small. In contrast to house dust mite allergy, mould allergy is no more in occupants of damp houses than dry homes, suggesting that individual sensitivity rather than current environmental exposure is the more influential determinant of sensitization to moulds. The principal house dust mite species in UK grows optimally at around 25°C and 80% relative humidity but less in the cold damp conditions associated with mould growth.

Not all comparisons of housing conditions show simple relationships between cold housing and poor health, but the evidence suggests an effect particularly on respiratory symptoms in children [Platt *et al.*, 1989]. One of the weaknesses of the studies is that people living in cold damp houses may differ in health and socio-economic background from those in drier, warmer homes.

3.6 Transmission of infections

In addition to the direct effects of cold temperature on the respiratory tract there are many indirect mechanisms involved in the transmission of infections in winter months in which indoor and outdoor environment interact (Table 3.2). It has been suggested, as discussed above, that increased transmission of infection may occur in winter because of increased survival of air-borne micro-organisms in lower humidities, while drying of nasal passages at low relative humidity may act to enhance the ability of micro-organisms to infect. High relative humidity, it is claimed, produces larger particles which reduce infectivity [Druett, 1967]. Another determinant might be the "enveloping" of viruses (including influenza and common cold) which enable the virus to survive outside the host for longer periods in winter when the relative humidity of the indoor (but not outdoor) air is very low. Adaptive strategies of micro-organisms are very important components of infectivity. For example, although cooler temperatures in the nasal passages retard the growth of those micro-organisms requiring warm environments, rhinoviruses which cause many common colds are adapted to grow better at the lower temperature found in the nasal passages [Toth and Blatteis, 1995]. In contrast to the effects of low humidity, studies of the viability of viruses in droplet spray at different relative humidities suggest that there is prolonged survival in more humid conditions [Buckland and Tyrrell, 1962]. Animal experiments point to 50% relative humidity as an optimum for limiting the infectivity of atomized influenza virus. Thus both high and low relative humidity may facilitate transmission of upper respiratory tract infection.

Table 3.2. Factors in cold-related infectivity

---

1.   Increased thermal capacity of humid, cold air
2.   Increased viral replication and transmissibility
3.   Survival of micro-organisms in indoor low humidity
4.   Microbial adaptation to the host environment
5.   Direct effect of cold air on bronchial reponsiveness
6.   Adverse effects of cold on immunological resistance
7.   Human adaptation to the microbial environment
8.   Seasonal human behaviour (e.g. overcrowding)

One of the clearest consequences of high population density is the increased transmission of infectious diseases, particularly respiratory diseases. From more recent discriminating analyses being applied to epidemiological investigations the independent effect of personal proximity i.e. overcrowding, among interrelated variables is emerging. The association of diseases such as chicken pox [Pollock and Golding, 1993] or the common cold [Jaakola and Heinonen, 1995] by droplet infection is a well-recognized hazard of winter behaviour in which closed living conditions occur more frequently in cold weather. Bronchitis, asthma and emphysema show highly significant associations with crowding in older groups, although as expected, they are also related to socio-economic differences [Kellett, 1993].

3.7 The elderly and vulnerable groups

Acute infectious respiratory diseases cause the highest mortality when they affect a vulnerable section of the population, such as elderly people already suffering from chronic disabling respiratory illness. Other factors increasing the risk of respiratory infection in old age include a reduced immune response, institutionalization promoting greater contact and transmission, and sometimes poor nutrition. Influenza is not more common in the elderly, but the complication rate, morbidity and mortality are much greater. The elderly are especially prone to "chronic crises" in winter and are susceptible to infectious diseases acquired because of debilitation [Collins, 1986]. The disease burden of upper respiratory tract infections in elderly people living at home is substantial. Overall, two thirds of elderly people with common colds and four fifths of those suffering from influenza or respiratory syncitial virus can be expected to develop lower respiratory tract complications [Nicholson *et al.*, 1997].

There are similarities in thermal comfort temperatures preferred by healthy young and elderly people [Collins and Hoinville, 1980] though the general tendency to assume that old people prefer warmer temperatures is basically correct. This is because most elderly are less active than the young and therefore require more external heat to maintain thermal balance and comfort. Nevertheless, households that include elderly people in the UK are 0.6° C cooler on average than other dwellings [English House Condition Survey, 1991] and 0.5° C cooler in the 1996 Survey.

Several studies have found an association between damp housing and respiratory disease, particularly wheeze in children, as has already been discussed. But there is often a poor correlation between reported wheezing and recorded consultations with general practitioners. There may be a strong element of bias in reporting respiratory conditions in children living in poor housing. In children less than 12 months old, the incidence of sudden infant death syndrome ('cot deaths') is greater in winter months. Many of these babies have had minor, often non-specific symptoms of respiratory illness and it has been suggested that a cold environment may have relevance in producing conditions where respiratory tract infections are more common. A significant correlation has been reported between minimum daily environmental temperatures and the incidence of 'cot deaths' four days later [Murphy and Campbell, 1987]. There are a number of other groups of people such as the chronic sick and disabled who, like the elderly and infants, are less mobile in their home environment and more at risk from the adverse effects of cold conditions.

3.8 Seasonal respiratory mortality

Table 3.3 Mean Excess Winter Death Index (EDWI) * by age and principal cause of death (1976-1983)

Cause	Mean EWDI* by age (y)			Mean EWD **
	45-64	65-74	75 plus	45 plus
Ischaemic heart disease	17	20	26	10912
Cerebrovascular disease	16	20	29	5317
Other cardio-vascular disease	21	26	33	5031
Respiratory disease	53	55	56	12878
Accidents and violence	10	18	37	857
All other	3	5	10	3631
All causes	12	18	27	38626

*   EWDI is the percentage excess of deaths in the four months of highest mortality (Dec - Mar) compared with the average of those in the preceding and following four months.
**   EWD is the total number of winter deaths less the average of those in the preceding and following four months
Source: Curwen [1997].

Respiratory mortality contributes a significant proportion of excess winter deaths and is influenced strongly by the degree of cold and the number of influenza deaths [Curwen, 1997]. In Table 3.3 the Excess Winter Death Index (which measures excess winter mortality as a percentage of non-winter mortality) is analysed in terms of cause of death and age. The index for all causes of death is more than twice as high at ages over 75 y as compared to those at 45-64 y. During the period 1976-83, circulatory disease caused 55% and respiratory disease including influenza 33% of all excess winter deaths among adults over 45 y. For non-winter deaths the proportions were 52% for circulatory and 16% for respiratory deaths. It is therefore on respiratory mortality that winter has the greatest proportional effect.

Over the period 1964-84, excess winter mortality declined noticeably while central heating in UK households rose from 13% to 66%. Part of the fall in respiratory mortality during this period is attributable to reduction in influenza epidemics. Although respiratory mortality has declined as house heating has improved (88% of UK housing now has some form of central heating according to the 1996 EHCS) excess cardiovascular mortality has not fallen correspondingly [Keatinge, Coleshaw and Holmes, 1989] (Fig. 3.2). These findings suggest that respiratory health may be more related to indoor cold, whereas outdoor cold may be more significant for cardiovascular morbidity. Nevertheless, there still exists a substantial excess respiratory disease mortality in the UK in winter [Curwen, 1997]. Minor upper respiratory tract infections may increase because of lower indoor relative humidity, as discussed above, but increased mortality from pneumonia suggests a direct adverse effect of cold stress.

As described in Chapter 1, more recent European studies have found that winter mortality increased to a greater extent for a given fall of temperature in regions with relatively warm rather than cold winters, in populations with cooler homes, and among people who wore less clothing [Eurowinter Group, 1997]. Evidence was obtained to link mortality with home heating independently of outdoor cold stress, and outdoor cold stress independently of home heating. A combination of warm outdoor clothing and warm housing (19-20° C) appears to prevent cold-related mortality in a community where outdoor temperatures are as low as -20° C, though respiratory but not cardiovascular mortality increases when outdoor temperatures are lower than -20° C [Donaldson *et al.*, 1998]. This survey in Siberia contrasts markedly with the winter increase in mortality that occurs in regions of the world with milder winters.

The issue of winter deaths arising from influenza is a complicated one, not least because of the imprecise certification of the disease. Even more problematical in assigning cause of death in winter mortality studies is the role of respiratory disease in cold-related mortality from arterial thrombosis. Winter mortality from stroke and ischaemic heart disease appears to be related to influenza epidemics as well as cold temperatures [Tillett *et al.*, 1983]. A large number of epidemiological studies have reported an association between various "inflammatory" factors and ischaemic heart disease. For example, the clotting factor fibrinogen is also an acute phase protein synthesized in increasing amounts during inflammatory processes e.g. respiratory infections. Raised plasma fibrinogen concentration is a recognised risk factor in arterial thrombosis

[Woodhouse *et al.*, 1994] which occurs during the winter. There are indications that short-term exposure to moderate cold can initiate a mild inflammatory reaction as well as a tendency for an increased state of hypercoagulability.

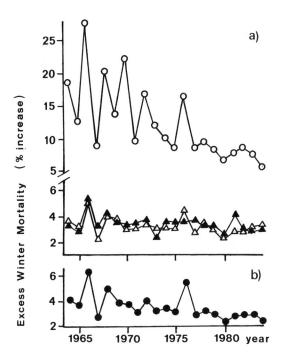

Fig. 3.2. Excess winter mortalities, adjusted for coldness of different winters in England and Wales for a) people aged 70-74 and b) for all ages. o , respiratory disease; ▲ , ischaemic heart disease; Δ , cerebrovascular disease; • , deaths from all causes. Source: Keatinge *et al.*, 1989.

3.9 Conclusion

Though of different intensity, outdoor and indoor cold temperatures both have the potential for promoting respiratory illnesses in the presence of respiratory pathogens. Severely cold outdoor temperatures may directly affect the natural defences of the respiratory system. Cold dwellings can have indirect effects through conditioning the humidity of the indoor environment and infectivity of micro-organisms. It should be recognised that many homes in the UK are damp and/or cold but that it is methodologically very difficult to show a definitive link between home temperatures and specific health outcomes. With regard to the effects of seasonal lower winter temperatures on respiratory disease it is outside temperatures that are measured and these do not usually distinguish between day and night, sudden changes, wind chill, humidity or atmospheric pollution. Interpretation of the causes of excess winter deaths is made difficult by blurring of the distinction between respiratory and cardiovascular deaths. The results of recent epidemiological studies show direct associations between a reduction in mortality indices and increased protective measures against the cold.

4

Cold stress, circulatory illness and the elderly

James Goodwin

4.1 Introduction

There is a widespread belief in the UK that large numbers of elderly people suffer death from hypothermia in the winter months. This is not borne out by the statistics which show rather that there is an excess of winter deaths due largely to respiratory and circulatory disease, such as heart attack and stroke [Curwen, 1997]. Several studies have now established that there is a relationship between these deaths and seasonal reductions in temperature [Bean and Mills, 1938; Dunnigan *et al.*, 1970; Bull, 1973; West and Lowe, 1976; Bainton *et al.*, 1977; Khaw, 1995] but the nature of this relationship is unclear. Many European countries with more severe winters than the UK suffer much lower winter mortality [Laake and Sverre, 1996] and though there has been a trend towards the greater use of central heating, the number of excess winter deaths caused by circulatory disease shows little decline [Keatinge *et al.*, 1989]. There is a trend towards fewer winter respiratory deaths [Curwen, 1997]. Recent studies examining the way in which populations are exposed to cold stress have indicated that outdoor excursions and the degree of cold protection involved, as well as the indoor climate, may play a role in the causation of winter deaths [Eurowinter Group, 1997].

This review will therefore examine the relationship between cold stress and the health of elderly people, with particular reference to circulatory disease. The issue of the relative importance of indoor and outdoor temperatures will be considered because there have been several studies which indicate that the level of indoor warmth alone provides only part of the possible protection against the effects of winter cold [Keatinge

1986; Coleshaw *et al.,* 1990; Donaldson *et al.,* 1997, 1998a and 1998b].
The role of physical activity will also be considered. Recent evidence
indicates that physical activity in the elderly may in fact be raised in the
winter and may possibly contribute to the effects of cold stress in
provoking higher blood pressure, a known risk factor for thrombotic
disease [Goodwin, 1999] .

4.2 Winter morbidity and mortality

EXCESS WINTER DEATHS

Ever since records have been kept in the UK, it has been known that the
number of deaths occurring in the winter is higher than at other times of
the year. As early as 1839 this is mentioned in the Second Annual Report
of the Registrar General and in the Third Annual Report, Farr [1841]
spends several pages discussing the effects of low temperatures on
mortality in London during the years 1838-41 [GRO, 1841]. McDowall
[1981] examined the period from 1841 to 1980 by analysing the trend in
winter deaths using periods of 10 years at a time. He calculated a
'Seasonality Ratio' by expressing the annual mortality rate for the
January to March quarter (known as the March quarter) as a ratio of the
yearly rate taken as 100 . He showed how the ratio for the March quarter
increased steadily from an average of 110 for the period 1841-50 to a
figure of 128 for the period 1921-30. This increase was largely due to
falling summer mortality. Since 1930 the ratio has been falling (for
example the 1979-88 average ratio was 114 [Curwen, 1991] and for
1986-90 it had reached 111 [Curwen, 1997]). Never the less, the problem
of winter deaths remains a serious one, as mentioned in Chapter 1, with
between 20 000 and 40 000 excess deaths occurring each winter in the
UK.

EXCESS WINTER DEATHS AND AGE

The number of excess winter deaths (EWD) is age related, those aged 65
and over being particularly at risk. Analysing monthly deaths in England
and Wales for the years 1962-67, Bull and Morton [1975] first showed
that the slope of regression of death rates (from myocardial infarction) on
ambient temperature becomes steeper with age (see Fig. 4.1).

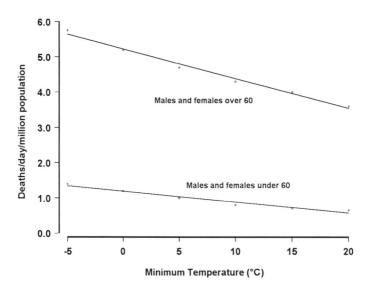

Fig. 4.1. Daily deaths from myocardial infarction in England and Wales, 1970-71 [after Bull and Morton 1975 reproduced with permission]

OPCS (Office of Population, Censuses and Surveys) raw data for 1983 are also very striking with regard to age. The average weekly total of deaths for January to March (1983) for those under 74 is 2071 compared to 3501 for those aged 75 and over, who represent only 12% of the total of all age groups [Alderson, 1985]. More recent data corroborate this disproportionate risk of mortality with respect to age. For example, if Seasonality Ratios are calculated for the very cold winter of 1984-85, for the age group 65-74 the ratio is 123 whereas for the age group 75+ it is 135. To further exemplify this point, a summary of average EWD data for 1970-1991 is shown in Table 4.1.

Table 4.1. Average excess winter death data , October-March 1970-91

	< 65	65-74	75+
Average Extra Deaths/month/°C	140	260	680

Source: Curwen, 1997

CAUSES OF WINTER DEATH

In the public mind, there is a strong assumption that large numbers of elderly people die from hypothermia in the winter. This assumption is not supported by the statistics. There are only between 500-600 recorded deaths per year where hypothermia is recorded on the death certificate [Collins, 1987] and in 95% of these cases, hypothermia is not registered as the underlying cause of death. More recent data show that deaths from hypothermia are still numerically insignificant [Curwen, 1997].

The principal causes of death in the winter months are in fact circulatory and respiratory disease. Mortality from respiratory disease accounts for about half of all excess cold related deaths with ischaemic heart disease and cerebrovascular disease the remainder [Curwen, 1997]. It has already been pointed out that the seasonality ratio for the UK appears to be falling. However, the true picture does not emerge until the causes of death in the winter are considered. Keatinge, Coleshaw and Holmes [1989] showed that excess deaths from respiratory disease did decline by 69% from 1964 to 1984 (even when summer mortality and temperature were taken into account) whereas coronary and cerebrovascular deaths did not fall significantly. Therefore, though it appears that there is a trend towards 'de-seasonality' in the UK, nested within this is the persistent problem of thrombotic deaths.

4.3 Temperature and excess winter deaths

Many studies have shown a relationship between environmental cold and deaths from circulatory disease [Rose, 1966; Bull, 1973; West and Lowe, 1976; Bainton et al., 1977; West, 1989; Khaw, 1995]. The 1987 OPCS regression model established environmental temperature as a strong predictor of excess winter deaths, showing that each degree Celsius by which the mean winter temperature varies is associated with about 8000 excess deaths. To illustrate this with an example, in the cold winter of 1984/85 with a mean temperature of 1.35°C below the average, there were about 10,500 *extra* EWD. A new monthly analysis [Curwen, 1997] confirmed that much of the annual total mortality is independently related to temperature (and also to influenza; *see* Chapter 3). It should be noted that the single measure of temperature used in Curwen's analysis is the average 24h dry bulb air temperature during the 4 winter months. As Dr Collins points out in the previous chapter, it therefore does not take account of 24 h variations in temperature, other meteorological factors such as wind-chill, or the differences between indoor and outdoor temperatures to which people are exposed.

LABORATORY STUDIES

Many laboratory studies have attempted to throw light on the possible physiological mechanisms operating in excess winter deaths. Two major areas of research have been the haematological and haemodynamic responses which may be occurring in the elderly when exposed to cold.

There are notable haematological changes which occur in elderly subjects in response to cold air exposure, which are not seen in young adults. Investigations exposing lightly clad young and old subjects to early morning cold (sufficient to cause significant changes in skin, but not core, temperature) have shown increases in major risk factors for arterial thrombosis. These changes include increases in erythrocyte and platelet count, packed cell volume and whole blood and plasma viscosity - all changes which might contribute to rapid increases in coronary and cerebral thrombosis in cold weather. Other changes such as raised cholesterol and fibrinogen concentrations may contribute to the slower components of the increased thrombosis [Keatinge et al., 1984; Coleshaw et al., 1990; Neild et al., 1994].

There are also age related differences in cardiovascular responses to the cold [LeBlanc et al., 1978; Wagner and Horvath, 1985; Budd et al., 1991; Collins and Goodwin, 1998]. For example, in a study in which elderly and young subjects were exposed to cold indoor temperatures, Collins et al. [1985] showed that changes in blood pressure (BP) occur significantly more slowly at first but are much more marked in the older subjects after two hours in still air at 6°C. A small rise in BP occurred in the elderly subjects alone at 12°C and no increase in either group at 15°C. It has also been shown that short term increases in BP (i.e. in less than two hours) are evoked by cold extremities and slight reductions in (0.2 to 0.4 °C) in core temperature [Collins et al., 1989]. Experiments on facial cooling show similar effects. Generally older people experience a greater increase in arterial BP and a less pronounced bradycardia than younger people with facial cooling (e.g. with a 6.0m/s air stream at 3.5°C). These are trigeminal reflexes and they appear to be enhanced by body cooling and diminished when the body is warmer, so that a colder elderly person is at more risk from hypertensive responses due to facial cooling than an elderly person who is warm [Collins et al., 1996].

Many cardiovascular reflexes show altered responsiveness as a function of age and this includes respiratory sinus arrhythmia, vagal baroreflexes, cardiopulmonary reflexes, reflex tachycardia, facial cooling bradycardia and cold pressor reflexes [Eckberg and Sleight, 1992]. In the elderly this is thought to be due to a decline in autonomic control [Duke *et al.*, 1976; Mancia *et al.*, 1980; Collins *et al.*, 1996]. The evidence for this includes comparison of heart rate variability in young and old subjects which has shown that spectral power in the low frequency ('vagal and sympathetic outflows') and high frequency ('vagal cardiac outflow') components are significantly lower in elderly people [Korkushko *et al.*, 1991]. The speed of response in many of these reflexes is therefore seen to decline with age and in the case of the control of BP, baroreflex response is diminished. This accounts for the uncompensated rise in arterial BP and the small reflex bradycardia with facial cooling in the elderly. On exposure to cold one therefore finds that vasoconstriction is slow and BP also rises inexorably but slowly with a smaller than expected fall in heart rate. The uncompensated levels of heart rate found in the elderly in the cold have been ascribed to increased sympathetic tone in the cardiac motoneurones or to the reduced withdrawal of vagal inhibition [Collins *et al.*, 1996].

FIELD AND CLINICAL STUDIES

Rose [1961] was the first to show the seasonal variation of blood pressure, though he was incorrect in describing the pattern as different to that of ischaemic heart disease (IHD) and mortality (which follow the reciprocal of air temperature - namely, a peak in January and February). Brennan *et al.* [1982] in a large study involving 1700 patients (age range 35-64) demonstrated a pressor effect (peaking in January and February and associated with cold outdoor temperatures) which was greater in older (55-64) and thinner patients than in younger (35-55) and more obese ones. Patients were borderline or mildly hypertensive (diastolic blood pressure (DBP) 90-109 mmHg). Blood pressure differences of 6-7 mmHg systolic pressure (SBP) and 3-4 mmHg DBP were shown for a 20°C seasonal change in maximum air temperature. Several other seasonal studies have been conducted but mostly using age groups other than the strictly elderly [Prineas *et al.*, 1980; Hata *et al.*, 1982; Jenner *et al.*, 1987; Knuiman, Hautvast and Zwiauer, 1988; Giaconi *et al.*, 1989]. It is difficult to coherently summarise these data and a number of issues remain unresolved, for example, the exact role of indoor and outdoor temperature, and the influence of medical condition and of age. In spite of the many confounding factors between different studies, one may observe an association between environmental temperature (seasonal variation) and blood pressure. Giaconi and Ghione [1992] attempted to quantify this relationship on the basis of a meta-analysis and concluded that a SBP and

DBP variation of about 7mmHg can be predicted for temperature differences of 20°C (i.e. 0.35 mmHg/°C). However, this analysis does not take account of the individual's BP status (e.g. hypertension) or age.

Field studies which have restricted themselves only to elderly subjects are few. Woodhouse, Khaw and Plummer [1993] took monthly measurements of BP, body temperature, indoor and outdoor temperatures and self-reported activity levels in 96 men and women aged 65-74 over a 13 month period. Regression analysis revealed highly significant seasonal differences in both SBP and DBP. Seasonal variation (mean (95% Confidence Intervals)) was 12.0 (9.5-14.3) mmHg for systolic and 5.5 (4.4-6.7) mmHg for DBP, both peaking in the winter. There were also significant seasonal differences in heart rate (2 bpm higher in the winter), body mass index (0.43 kg/m^2 higher in the winter) and certain outdoor physical activities (e.g. walking; 1.3 more walks per week in the summer). Both living room temperature and mean daily outdoor temperature showed strong and independent inverse relationships with blood pressure. Indoor temperature showed the strongest relationship: a 1°C decrease in living room temperature was associated with rises of 1.3 mmHg SBP and 0.6 mmHg DBP. Interestingly, the seasonal differences in BP are greater than those in previous studies. One reason for this may be the presence of an order effect inflicted from the experimental design. There was also evidence of the so-called 'white-coat' (i.e. physician related) effect on BP. 'Casual' or 'home' spot-readings of BP are known to be susceptible to this effect and are less reproducible and more varied than ambulatory BP data [Mansoor et al., 1994]. Another criticism of the Woodhouse study is that no temperature control was possible over the one-off home BP readings, though each subject was measured under similar conditions in their own home from one reading to another. This issue is important because it has been shown that there is a marked effect of room temperature on BP, with higher temperatures strongly associated with lower blood pressures irrespective of season. In fact, standardisation of room temperature has been found to remove seasonal effects [Heller et al., 1978]. One other notable study deserves mention. Stout and Crawford [1991] carried out a large study involving 100 subjects aged 75-82 but found no evidence of seasonal variation of blood pressure. It is not readily apparent why this finding should be disparate to that of other studies.

In relation to blood pressure, season and age, the position may be summarised as follows. Most studies investigating effect by season have measured heterogeneous populations in terms of age and BP status and have disclosed higher blood pressures in the winter. Only two major studies [Woodhouse et al., 1993; Stout and Crawford 1991] have carefully restricted their populations to over 65 years and their findings are equivocal. No field studies have taken contemporaneous readings from a young control group. There is therefore a need for a field or clinical study in which these issues are addressed.

4.4 Cold stress and behavioural risk

Notwithstanding the clear negative relationship between ambient temperature and the incidence of circulatory disease, there are anomalies in the literature which require explanation. Briefly stated, seasonal mortality in the colder northern European countries is *lower* than in those with milder winters [McKee, 1989; Curwen, 1991; Laake and Sverre, 1996] and secondly, high indoor temperatures alone do not appear in themselves to offer protection against winter mortality [Alderson, 1985; Keatinge, 1986; Curwen, 1997].

The important role of adequate indoor warmth in the prevention of winter illness in the elderly has been emphasised much in the literature [Watts, 1972; Collins and Hoinville, 1980; Collins, 1987; Herity *et al.*, 1991; Khaw, 1995]. A large proportion of the elderly, the most vulnerable section of the UK population, have no central heating in their homes or cannot afford to use it fully [Collins, 1986; OPCS, 1987; Keatinge, Coleshaw and Holmes 1989; Boardman, 1991]. Substantial numbers of the less well off may therefore be exposed to low indoor temperatures in the winter (there are 2.5 million homes where the warmest temperature is less than 16°C [DoE, 1996]). Poor insulation will exacerbate the situation [Fox *et al.*, 1973; Collins and Hoinville, 1980; Hunt and Gidman, 1982]. Further, the heating of only one room may lead to the occupants experiencing fluctuations of temperature as they move from one room to another, much in the same way that a person venturing outside may experience. In relation to the influence of indoor temperatures on cardiovascular responses, studies have shown that 15°C appears to be the threshold temperature for pressor effects in elderly people and therefore this would be a minimum level at which elderly people should live in their homes [Collins *et al.*, 1985; Collins, 1986].

However, it is becoming increasingly clear that indoor warmth may be a necessary but insufficient condition for the maintenance of the health of elderly people in the winter. There is evidence to show that the excursional exposures of everyday life may play a part in increased winter mortality in the elderly [Keatinge, 1986; Donaldson *et al.*, 1997] and that the mechanisms described above in relation to cold (haemoconcentration and hypertension) may indeed be extant in the elderly who are subject to episodic cold stress [Donaldson *et al.*, 1997]. Studies examining the behavioural patterns of people living in different European regions appear to show that the avoidance of cold stress (restricting outdoor cold excursions, wearing warmer clothing, increasing activity when outdoors) is associated with low indices of cold-related mortality in the countries with the most severe winters [Eurowinter Group, 1997; Donaldson et al., 1998a and 1998b]. Conversely, it may therefore be the case that 'high risk' behaviour (that is, making excursions into the cold outside air, with lower clothing insulation than is necessary to afford adequate protection) coupled with inadequate indoor heating is contributing to the excess winter mortality in the UK. Previous studies have shown that many old people have to move repeatedly between warm and cold rooms in the winter [Collins, 1986] and that emerging from a cold dwelling into the cold may produce greater cardiovascular strain than leaving a warm dwelling [Collins et al., 1989].

More recent prospective data corroborate the idea that excursional behaviour may predispose the elderly to the ill effects of cold in the winter [Goodwin, 1999]. In an ambulatory study of 25 healthy elderly subjects (age range 70-82 years) in two successive winters and summers, it was found that though the total number of excursions was greater in the summer than in the winter, the hypothesis that the average time spent outdoors per excursion was the same for winter and summer held in all cases. Thus there appeared to be a transfer of unmodified excursional behaviour from the summer to the winter. In this study, the mean (±sd) indoor living room (dry bulb) winter temperature was 18.5±2.1 °C and the corresponding mean daytime outdoor air (dry bulb) temperature (and range) was 5.8°C (-4.6 to 9.7°C), giving an indication of the temperature gradient experienced during the outdoor excursions. Ambulatory daytime blood pressure (e.g. SBP (mean±sd)) was significantly higher ($p \leq 0.01$) in the winter (142±14 mmHg) than in the summer (134±14 mmHg) and a regression model for this elderly group predicted a rise in SBP of 0.47 mmHg/°C, slightly larger than the 0.35 mmHg/°C indicated by Giaconi and Ghione [1992] for younger age groups. On average, in this group of fit elderly persons, there were about 5 excursions per day of between 20 and 30 mins duration. A much larger prospective study of the elderly population as a whole would be required to provide more representative data, however.

4.5 Role of physical activity

Explanations of the seasonal, i.e. winter, increases in blood pressure have centred on the effects of ambient temperature. However, other studies

[Erikssen and Rodahl, 1979; Sandvik *et al.*, 1993; Mundal *et al.*, 1997] have examined the role of physical activity, which will now be discussed.

SEASONAL DIFFERENCES IN ACTIVITY

There is surprisingly little evidence as to the existence of seasonal variation in physical activity in either young or elderly subjects. This may be accounted for largely by the practical difficulties of measuring activity levels in the general population [Wilson, Paffenbarger and Morris [1986]. Physical activity is a complex behaviour with several interrelated dimensions. There are therefore numerous methods of assessing activity but the most widely used method of collecting data on physical activity is by questionnaire or activity diary, for mostly practical reasons [Lamb and Brodie, 1990; Voorips *et al.*, 1991].

The first major study to examine physical activity by season and by age was the Framingham Offspring Study [Dannenberg *et al.*, 1989] which surveyed over 3000 subjects aged 17 to 77 by questionnaire. Substantial seasonal variation was found, subjects being more active in the summer months. An inverse relationship was found between level of activity and cardiovascular risk factors, as in other studies [Morris, Everitt and Pollard, 1980; Pfaffenbarger and Hyde, 1984; Kannel *et al.*, 1985]. A second major study (16 000 subjects) corroborated the Framingham findings: Uitenbroek [1993] found considerable seasonal variation in both outdoor and indoor activities with older respondents showing more seasonal variation than younger ones. Activity (defined as exercise periods of more than 20 minutes) peaked in the summer in both groups. Similarly, in a field study designed to measure seasonal variation in blood pressure in the elderly, Woodhouse Khaw and Plummer [1993] showed an increase in self-reported activity in the summer. However, in a study in which activity was measured objectively (by means of an ambulatory accelerometer carried on the non-dominant wrist) elderly subjects were shown generally to be more active in the winter than in the summer [Goodwin, 1999]. This reported increase in activity may be a behavioural adaptation to the cold stimulus of outside temperatures and colder indoor temperatures. A time series analysis of the activity data produced a model the parameters of which also varied significantly differently according to season (winter and summer). Younger subjects (age range 18-30) were consistently shown to be more active than the elderly ones in both seasons. Further research is required to clarify the position since there appears to be a dichotomy between those studies using subjective self-report data and those using more objective measures.

PHYSICAL ACTIVITY AND BLOOD PRESSURE

The relationship between seasonal variation in physical activity and blood pressure is an interesting one and has been little studied. Blood pressure as a measure is highly variable. The time course of its variability ranges from a few seconds or minutes (short-term) to 24 h (long term) and to one year (seasonal). At any point in time it is determined by endogenous and exogenous factors, such as level or intensity of physical activity, environmental factors (e.g. temperature) and contextual influences (psychosocial factors, e.g. stress). Equally such factors may exert a long term influence. Therefore, in the literature there are typically two issues under discussion. One issue is the long term health related impact of exercise or individual regime of physical activity on resting blood pressure. The consensus position would be that increased habitual physical activity reduces resting blood pressure [Choquette and Ferguson, 1973; Hurley et al., 1988; Friedewald and Spence, 1990; Pfaffenbarger et al., 1993]. Should physical activity increase in the summer months, then the predicted effect would be a reduction in resting blood pressure and seasonal differences could be explained on this basis. Mundal et al. [1997] have in fact argued that the seasonal increases in blood pressure which are ascribed to the effects of winter cold may be explained on the basis of winter reductions in physical fitness in middle aged men. The second issue to which authors refer is the concurrent effect of physical activity on blood pressure. The general observation of directly proportionate increases in systemic blood pressure with increased whole body activity is well established in physiology [e.g. Åstrand et al., 1965], the exercise effect persisting with physical demand and subsiding during recovery. The impact of these short term changes may be seen where BP and activity are quantitatively measured simultaneously and the relationship described as a 24 h or circadian rhythm. Data from two studies are shown in Fig. 4.2 [Collins et al., 1995 and Goodwin, 1999].

Comparison of the BP circadian rhythm of resting subjects with that of freely active subjects clearly shows the amplitude effect of quiescent behaviour. Resting subjects here were largely sedentary throughout the study which took place in a controlled environment (ambient temperature 21°C dry bulb, 50% relative humidity, air velocity 0.1-0.2 m/s) to which subjects were restricted for 48h. On the other hand, the BP profile of the active subjects results from ambulatory BP data taken from subjects undertaking their normal course of activities of daily living in their own homes.

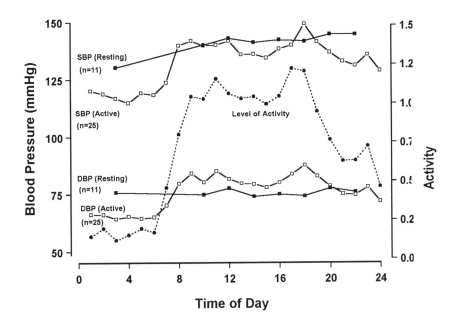

Fig. 4.2. Elderly blood pressure in conditions of activity (level of activity shown as transformed accelerometer counts) and rest (subjects sedentary) over a 24 hour winter period.

A second observation would be the close apparent coincidence of activity level and blood pressure which shows a typical early morning rise, daytime plateau, evening decrease and nadir during the night. This is entirely consistent with previous studies which strongly suggest that most, if not all, of the daily rhythm in blood pressure may be ascribed to the effects of the activities of daily living. The normal 24h rhythm of blood pressure is known to have a large exogenous component and much of the diurnal rhythmicity is imposed by the daily schedule of movement and activity, changes in body posture and emotional and social interactions [Harshfield, 1992; Pickering *et al.*, 1992; Shapiro and Goldstein, 1998]. [van Egeren, 1992; Baumgart, 1992]. The debate as to whether the BP rhythm is endogenously or exogenously driven seems to be moving towards the view that it is predominantly determined by the cycle of physical activity or arousal [Pickering *et al.*, 1992].

Though studies have shown seasonal BP variation in middle aged and elderly people (BP higher in the winter) few studies publish circadian data to show if there are in fact time of day effects on BP which vary between winter and summer. Where these studies have been done (e.g. the multicentre HARVEST Study [Winnicki *et al.*, 1996]) data is generally not reported on the elderly population. In the HARVEST Study 24h ambulatory BP was reported on 46 subjects aged 18 to 45 years, which showed that SBP in the winter was significantly higher than in the summer between the hours of 0900h to 2100h but not during the night. There was no seasonal difference in DBP or heart rate. In elderly subjects, a similar pattern has been reported for SBP (significantly higher in the winter during the day (0900h to 1200h) but no significant difference at night) but unlike the HARVEST study, these differences were accompanied by significant increases in both DBP and heart rate [Goodwin, 1999]. It therefore appears that at certain times of the 24 h cycle, cardiovascular strain in the elderly is significantly raised in the winter compared to the summer and that these time of day effects correspond to the greatest increases in activity (Fig. 4.2).

IMPLICATIONS FOR HEALTH

From 1985 onwards a series of studies confirmed that there exists a circadian variation of thrombotic illness and that its acute onset may be associated with concurrent elevations of blood pressure, particularly during the morning and early evening rise [Muller *et al.*, 1985; Muller, Tofler and Stone, 1989; Tsementzis *et al.*, 1985; Robertson *et al.*, 1986; Becker and Corrao, 1989; Pickering 1990]. The consensus by 1990 was that the early morning rise in BP most probably represents a synergy between the endogenous rhythm and the effects of awakening and activity. Other recent evidence has shown that there is a strong association between 'external triggers' and the onset of sudden cardiac death, beyond that which may be attributed to chance alone [The Onset Study, Willich *et al.*, 1992; and The 'TRIMM' Study, Mittleman *et al.*, 1993]. Physical exertion was shown to be one of the most potent 'triggers' and it has been shown by ECG analysis that transient ischaemic episodes and increased myocardial demand correspond to the morning onset of activity [Parker *et al.*, 1994]. Though direct causal evidence has yet to be brought forward, the hypothesis that thrombotic events may be triggered by early morning activity is gaining ground [Purcell *et al.*, 1992; Willich, 1997]. The findings reported above - that there is increased blood pressure and heart rate in the elderly in the winter compared to the summer, at precisely the times of day (early morning and early evening) which correspond to the times of greatest risk of the acute onset of thrombotic disease - therefore appear to have important implications for preventing the primary occurrence or recurrence of acute thrombotic events in the elderly. Physical activity generally and a high level of physical fitness have been shown to have beneficial outcomes in relation to health, notably reducing the risk of mortality from cardiac heart disease in the long term [Pfaffenbarger *et al.*, 1993]. Therefore, advice on lifestyle has to be highly specific: it would appear that it is advantageous for elderly persons to be habitually active or physically fit but that they should avoid vigorous activity at certain times of the day, more especially so in the winter and they should moderate their outdoor excursions into the cold.

Other studies have shown that prothrombotic processes such as increased platelet adhesion, blood viscosity and platelet aggregability occur during the morning hours after rising and that the magnitude of these increases are associated with the level of activity [Tofler *et al.*, 1987]. Unfortunately from a morning risk point of view, fibrinolytic factors in the blood have a circadian trough which occurs in the early morning [Muller, Tofler and Stone, 1989] reducing the possible mitigation of thrombotic tendencies. These kinds of haematological changes are known, as indicated earlier, to be exacerbated by exposure to early morning cold [Keatinge *et al.*, 1986].

4.6 Conclusions

Excess winter deaths in the UK have showed some decline over the past 20 years (the so-called 'de-seasonality') but in spite of the increased use of central heating the problem of deaths due to circulatory disease has persisted, particularly in the elderly. Indoor temperatures play an important role in mitigating the effects of winter cold on the elderly population and a multidisciplinary approach in enabling the elderly to maintain adequate levels of indoor warmth should be taken. However, there is additional evidence that inadequately protected exposure to outside cold contributes to winter mortality. The effects of cold stress on the elderly may be reduced by restricting the number and duration of outside excursions into cold air and by improving individual protective measures against the cold. The risk of the onset of acute circulatory disease appears to be greater at certain times of day than at others, notably during early morning activity. Though the risk of thrombotic illness is reduced in those who are fit and active, the rate of sudden death during brisk physical activity is higher than the average rate in sedentary men [Willich, 1997] and therefore on rising intense activity would appear to be contra-indicated in the elderly, particularly in the winter.

5

Dust mite allergens, indoor humidity and asthma

Stirling Howieson and Alan Lawson

5.1 The rising incidence in asthma

The rise in the prevalence of asthma has been a point of contention over recent years for researchers investigating the topic. Attributable factors ranging from pollution, changes in diet, a more hygienic lifestyle, the prophylactic use of antibiotics and hygrothermal changes in the domestic environment have all been implicated as playing a major role, as has over-reporting due to a greater public awareness of health related issues.

Previous studies [Burr 1989, Rona 1995, Omran 1996] suggest that the prevalence of asthma has risen and cannot be wholly explained by greater awareness (at least part of the increase is real). In a study of asthma in the Scottish highlands [Austin 1994], and in a national survey [Strachan 1994], both concluded that macro-environmental factors, such as outdoor pollution and climate, although known triggers of asthma [Andrae 1988], were unlikely to be responsible for the rise in prevalence of asthma. Both studies reported similar rates between natives and migrants, in rural and urban areas, although the severity and frequency of attacks were higher in urban areas - this trend also paralleled with decreasing socio-economic status.

Associations between house dust allergens and the house dust mite were made as long ago as 1928 [Dekker 1928], re-iterated later by [Voorhorst 1967] when the species of mite dermatophagoides was identified as a source of allergen capable of inducing allergic reactions. A further study [Korsgaard 1983] demonstrated that there was a clear dose-response relationship related to exposure to allergenic particles, furthermore, in a review of literature on the subject of mites and their allergens [Sporik 1992] it was indicated that the evidence strongly suggested that a causal relationship exists between exposure to mite allergens and asthma.

5.2 Housing and health

The health implications of living in damp homes have been studied in several epidemiological studies [Burr 1981, Burr 1989, Martin 1987, Platt 1989]. The results have demonstrated that home dampness is strongly associated with respiratory conditions with studies [Brunekreef 1989, Dekker 1991, Strachan 1988, Strachan 1989, Williamson 1997] linking damp housing specifically to asthma. A study of Scottish housing stock established that 25% of all dwellings in Scotland suffer from problems of dampness or condensation [Scottish House Conditions Survey - SHCS 1996].

Between 1991 and 1996 the percentage of dwellings in Scotland with double-glazing increased from 36% to 62% [SHCS 1996]. Although some of this increase will be accounted for by new build (in reality less than 5% of the total increase) over 550,000 dwellings had replacement units fitted. This compares with only a 3% increase in wall insulation. In the public sector these windows were primarily PVCu units which would have been replacing ageing timber or notoriously draughty steel units [SHCS 1996]. The effect of such a programme could be twofold:

(i)   to reduce the fortuitous level of air changes which would have previously facilitated water vapour dispersion;

(ii)   to reduce the sacrificial condensate effect which single glazing provides.

In dwellings where double-glazing is the sole improvement measure (i.e. no complementary upgrading of insulation or commensurate ventilation facilities e.g. extract fans, trickle vents etc.) it is likely that internal water vapour pressures would rise, especially where internal clothes drying is common. Although the standard advice disseminated by public sector landlords has been "turn up your heating and open your windows" [Scottish Consumer Council 1978], increasing air change rates by opening windows simply allows expensive heat to escape. It appears that the choice is between draughty windows that render the dwelling hard to heat, or double glazing, insulation, and central heating with a possible resultant increase in internal water vapour pressures and air temperatures, which provide the ideal conditions for dust mite proliferation. Part of the solution however may be at hand with the new generation of heat recovery domestic extract fans which will allow relatively high air change rates while maintaining the effective air change rate (i.e. ventilation heat loss) below 25% of the actual figure. The use of such a ventilation facility may have a further advantage as it may also remove a significant proportion of any airborne allergens.

5.3 Hygrothermal changes in the domestic environment

The domestic environment has changed rapidly over the past century, with a continuing downward trend in internal volumes. The energy efficiency drive, resulting from the fuel crisis in the mid 70's further exacerbated the problem encouraging tighter design and reduced air change rates. The introduction of vapour barriers into building skins effectively enclosed the occupants in a "plastic bag". Social activities have also evolved over this period; showers have become commonplace, as has indoor clothes drying.

These problems may have been further aggravated by the growing gulf in economic prosperity between rich and poor. Since 1971 the percentage of the population below half-average income has increased from 11% to 21% [Office for National Statistics 1996]. Many families in the lower income deciles find themselves in "fuel poverty", a condition which was described as an inability to adequately heat the home [Lowry 1989]. The people that are trapped in this social grouping are usually the unemployed, the chronically sick or the elderly, who may spend much of each day at home, therefore requiring to apply heat for longer periods. Fuel poverty will normally result in a dwelling being at best partially heated and typically, moist air will migrate from the kitchen/bathroom and condense in thermally isolated bedrooms.

5.4 The house dust mite

Research into the flora and fauna of house dust has been developing over the past two centuries, as have the health issues related to this field of study. The house dust mite has been of particular interest to many researchers over this time, and the specific mite family "Pyroglyphidae" are now thought to be of particular importance [Turos 1979]. The genus Dermatophagoides and especially the two species Dermatophagoides Pteronyssinus [Troussart 1897], and Dermatophagoides Farinae [Wressell 1974], the most abundant mites in Western Europe (including UK) and North America respectively [Blythe 1979, Burgess 1993], are now known to be a key factor in the development of allergic disease [Maunsell 1968, Miyamoto 1968]. An investigation into the rise in the prevalence of asthma in Papau New Gineau [Dowse 1985], demonstrated the house dust mite was a significant feature in the pathogenesis of the condition - 91% of the asthmatics had significant skin test reactions to extracts of Dermatophagoides mites and had grossly elevated serum-IgE antibody levels to the same species. 89% of the asthma sufferers questioned thought that their asthma had begun in the period subsequent to the acquisition of inexpensive cotton blankets (which they had never used before). The blankets are now used every night, are rarely washed and have provided a new environment for dust mite colonies to inhabit. Compared with a neighbouring tribe with a similar environment and lifestyle but with no access to cotton blankets, the incidence of asthma was 36 times greater.

The ideal conditions for mites to proliferate is at a temperature of 25°C and a relative humidity of 80% [Arlian 1989, Hallas 1990]. A high humidity is very important to the survival of these creatures as most of their water is gained from the atmosphere by osmosis [Hallas 1990]. The mites live in an atmosphere where no liquid water exists and moisture balance is critical to their survival. Under ideal conditions the life span of a mite is approximately three months [Wharton 1976]. Mite numbers vary seasonally, rising and falling in accordance with the humidity cycle within the house [Spieksma 1967, Platts-Mills 1989]. The highest numbers are experienced in middle to late summer [Wharton 1976], when relative humidity is usually highest [van Bronswijk 1971].

The main food of mites is human skin, thus heavily used soft furnishings, especially mattresses and bedding provide a perfect environment for the development of mite colonies [Sesay 1972, van Bronswijk 1973]. The flakes of skin absorb moisture from the atmosphere and are colonised by a yeast mould on the surface [Whitrow 1993]. The yeast causes the scales of skin to swell [Whitrow 1993], moistening and softening them to aid digestion. As well as skin, research [Douglas 1989] established that fungi of the genus Aspergillus to be of particular nutritional importance to mites. Aspergillus is one of the most common moulds found in damp public sector dwellings in Scotland. This adds further importance to the role of humidity in the life of the house dust mite, as moulds generally require a relative humidity of 65% or greater to exist [Gravesen 1979, Hart 1990].

HOUSE DUST MITE ALLERGENS

It was work in the late 1960's [Voorhorst 1967, Voorhorst 1969] that identified house dust mites as a potent source of house dust allergens. Further research [Tovey 1981] implicated the excretion products of the mites be a highly concentrated source of allergenic material. The work determined that mite faeces are minute particles, roughly spherical in shape and range from 10 - 40 μm in diameter, in direct proportion to the size of the mites producing them, which vary from 170-350 μm. The excretion products of Dermatophagoides Pteronyssinus were given the name Der PI. Another allergen produced by this mite has also been identified and named Der PII. These allergens are distinct proteins found in the faeces and in/on the bodies of the mites themselves [Wharton 1976].

The World Health Organisation (WHO) has suggested safety limits for exposure to dust mites and their allergens of 100 mites (or 2 μg Der PI) per gram of house dust [Platts-Mills 1989]. Above this level prolonged exposure increases the risk of sensitisation and the production of mite specific antibodies. If a level of 500 mites (or 10 μg Der PI) per gram of dust is exceeded there is an increased risk of a severe allergic reaction.

Although when inhaled most of these allergenic particles are trapped in the mucous membranes of the nose and throat, it was discovered that individual faecal pellets contain very high concentrations of allergen, and any particles that reach the lungs can trigger allergic reactions in susceptible individuals [Tovey 1981].

5.5 Asthma – pathology, sensitizers and triggers of asthma

Asthma is a disease that affects the lungs and the airways through which we breathe. During an asthma attack, the walls of the airways become inflamed, and the mucous membrane covering the walls becomes swollen with fluid. Viscous mucous fills the remaining air gap, making it difficult to breathe.

It is widely accepted that asthma attacks can be triggered by a number of different stimuli including cold weather, exercise, mould, pets or dust. A minority of asthmatics suffer from the condition of non-allergic asthma, where patients display typical symptoms of asthma but the disease does not appear to be provoked by any allergen in particular. This is a more common condition when asthma develops later in life [Whitrow 1993, Greener 1997].

In most sufferers, and especially children, asthma is an allergic disease provoked by a known foreign invader entering the body. An allergy can be defined as an exceptional response by the body's immune system, to a normally harmless substance or antigen. An antigen is described as any foreign protein that enters the body that induces the production of antibodies - in the case of allergic individuals, IgE antibodies are produced and any foreign proteins stimulating this specific response are called allergens. Antibodies are disease-killing cells which the human body produces to attack and destroy any antigens. When anything enters the body the immune system attacks and tries to kill the invader.

In allergic individuals, when an allergen enters the body the immune system tries to remove them by mobilising the appropriate cells. Whilst this process is occurring, the individuals can experience soreness, swelling and sometimes fever. These symptoms are due to the activities of the cells and chemicals within the body, in attacking and isolating the antigen. Symptoms specific to asthma include wheezing, shortness of breath, tightness of the chest, and cough (Asthma & the immune system review [Barnes 1992]).

Initially, most asthmatics are sensitised to one particular trigger. Once this condition has developed, it is usually the case that attacks can be precipitated by a number of other substances, as well as the initial agent [Rees 1995]. The most common allergic trigger in the UK is the house dust mite - it is estimated that between 50 - 75 % of allergic asthmatics react to extracts of mite allergens [Colloff 1992]. Other conditions induced by allergen exposure include eczema, rhinitis and hay fever.

5.6 Prevention measures

Due to the growth in interest in the house dust mite and its associated allergies, over the years a variety of different theories have been introduced (reviewed [Colloff 1992]), with the same desired outcome - to kill mite colonies. Measures applied and tested have included vacuuming, poisoning (acaricidal treatment), freezing with liquid nitrogen, baking mites and Der P1 with steam and increased ventilation to reduce humidity.

Vacuuming has proved ineffective [Mulla 1975, Carswell 1982, van Bronswijk 1985] as the mites are able to cling to carpets and other furnishings by the use of "suckers" on their legs [Wharton 1976]. A further problem with this method, [Sly 1985] is that the actual act of vacuuming stirs up the allergen, making the particles airborne, therefore increasing the chance of inhalation and subsequently triggering asthma attacks. On the positive side however, manufacturers are now becoming aware of the problems that dust mites caused and ranges of more powerful, improved filtering devices are starting to appear on the market.

Substances which, when applied to dust mites kill them via chemical action, are termed acaricides [Morrow Brown 1991]. The main drawback of the use of acaricidal products, which range from caffeine [Russell 1991], to highly potent organophosphates (toxic to humans) [Mitchell 1985], is that although they can be successful in killing mites and breaking down mite allergens, many of the products can themselves be dangerous. Furthermore the results from certain trials [Colloff 1990, Platts-Mills 1992] indicated that the method of application of the acaricide, and other local factors - such as depth and type of carpet, quantity of dust etc. - may affect the efficacy of the treatment.

The freezing of dust mites has been demonstrated to be a very effective method of eradication [Colloff 1986, Dorward 1988] - when combined with rigorous vacuuming after application. The method is commercially available. However it requires an expert operator who is knowledgeable in the safety aspects of handling and application.

At the other end of temperature extremes, steam has been used to kill and eradicate completely colonies of dust mites within bedding and carpets. In addition, the use of steam at high pressure also reduced the concentration of Der P1allergen in carpets by a substantial percentage in trials already completed [Colloff 1995].

The use of ventilation systems for the control of house dust mite numbers has been evaluated in several studies [Korsgaard 1983, Korsgaard 1991, Harving 1988, McIntyre 1993a, McIntyre 1993b]. The World Health Organisation set a figure for absolute humidity of 7g/kg as the limiting factor for the growth of colonies [Platts-Mills 1989]. Below this level numbers of mites begin to fall, due to direct desiccation of the mites themselves plus the dehydration of the skin scales on which they feed. The allergens produced by the house dust mites, however, do not decay naturally and have been shown to be stable for at least 4 years [Kort 1994]. A reservoir of antigen can thus be established and maintained over a period of time. Killing the mites themselves may lead to no improvement in symptoms, as the allergens will remain and a combined prevention/avoidance regime may prove to be of most benefit to the asthma sufferer. The use of ventilation should be combined with heat recovery to offset the considerable heat losses that would result from higher air change rates.

5.7 Meta-analysis

A recent meta-analysis published in the British Medical Journal concluded that current physical mite eradication techniques are ineffective and would therefore offer little benefit to the asthma sufferers. However, of the 24 completed studies examined in the paper, only 6 succeeded in reducing dust mite/allergen levels and in these studies a positive response was demonstrated.

The review undertaken identified 429 papers that held some relevance in the field of asthma and house dust mites [Gotzsche 1998]. When reviewed against their chosen criteria, 24 completed studies were examined in detail; 18 of these trials failed to reduce dust mite levels – either through inappropriate methods or improper application. Of the 6 remaining studies a variety of remedial procedures were implemented and assessed.

One study [Walshaw 1986] selected 50 adult asthmatics, with strongly positive skin prick tests to house dust mite extracts, and attacked bedroom allergen reservoirs. Mattresses were encased, bedding renewed and a weekly cleaning programme advocated to participants. If possible, carpets and all unnecessary soft furnishings removed. The study concluded that successful house dust eradication procedures are possible and that patients allergic to the house dust mite appeared to have both subjective and objective improvements in their asthma. There is also a suggestion that a longer trial including the use of mattress protectors would be of interest.

A similar project [Dorward 1988], which included the initial killing of mites with liquid nitrogen, also demonstrated improvements in 21 adult asthmatic patients and indicates the initial killing of mites should be included as an essential part of any future study. The study however did not measure Der PI allergen levels or assess whether liquid nitrogen had any effect on the existing reservoirs.

A different study [Enhert 1993] split 24 asthmatic children into 3 groups where one group used an acaricide on bedroom carpets and mattresses, a second used a placebo acaracide and the third utilised an encasing regime. Only the third strategy, which involved encasing of mattresses, duvets and pillows produced any significant reduction in dust mite allergen levels and indicated that this had some positive results in terms of reduced bronchial hyperreactivity and increased PC_{20}. Similar methods were used and results found in another trial [Carswell 1996], although it was indicated that the techniques may be of more benefit to highly mite sensitive asthmatics.

One further trial [Warner 1993] supplied ionisers for 20 asthmatic children to be used in living rooms during the day and bedrooms at night. Although the amount of airborne allergen was reduced there was no improvement in the participants' symptoms. Suggestions for this result identified that the ionisers were being run when disturbance to reservoirs was at a minimum and they may not have been so effective if more allergen was airborne. Also questioned is the importance of prolonged low exposure (airborne allergens) versus short duration high exposure (sleeping in high concentrations of allergen).

The final study that reduced allergen levels argued that the education of any participants in dust mite avoidance techniques was vital for a successful prevention regime [Huss 1992]. The sample of 52 adult asthmatics was split into 2 groups where one group received conventional instructions (counselling and written instruction) and the other received conventional instruction plus 22 minute of interactive computer-assisted instruction. Both groups lowered allergen levels on the mattresses over the 12 week monitoring period, with only the latter group achieving lower dust mite levels on bedroom carpets and floors. Bronchodilator use was reduced in the computer-assisted group, however no other symptoms improved.

5.8 An interventionist double-blind placebo controlled trial

In October 1997, the Centre for Environmental Design and Research (CEDAR) at the University of Strathclyde initiated a case based interventionist study involving approximately 70 asthmatics. The study is being conducted over a 30 month period to investigate the role of the house dust mite and its relationship with asthma within the domestic environment. The data collection programme was initiated in October 1998 with each dwelling monitored for a 6 month period before heat recovery fans were installed. In addition, a subset of homes were supplied with new duvets and pillows, with carpets being thoroughly steam cleaned and mattresses protected by breathable covers. These measures focused on removing any live mite colonies and denaturing allergen reservoirs deposited by the mites. The 18 months of monitoring (currently in progress) following the remedial measures will indicate whether the changes have had any significant effect on asthmatic symptoms when compared with two separate control groups.

The hypothesis that the current study team has adopted hinges on the need for a more comprehensive dust mite prevention and avoidance regime. The study has been designed to monitor and assess whether the 'real' rise in the condition of asthma and asthmatic symptoms is related to dust mite allergens, which have become ubiquitous within the home. More specifically, it aims to identify whether the house dust mite is a major contributory factor to the development of the asthmatic state, the ignition of asthmatic symptoms and sustainment of the disease within the housing stock of West Central Scotland. To fully investigate this hypothesis requires an understanding of the built environment, the physiology of the disease and production, properties and effects of dust mite allergens.

In order to encompass an understanding of each of these knowledge bases, a multi-disciplinary project team has been assembled. This includes researchers from the field of architecture, environmental engineering, public health, clinical medicine, clinical immunology, statistics and health economics.

GENERATING THE SAMPLE

From the outset the study identified North Lanarkshire Council as a suitable local authority partner, who were active in the field of housing and health and had expressed an interest in supporting applied research. The authority was however keen that the research was undertaken in specific catchment areas which had higher levels of fuel poverty.

All public primary and secondary schools within these areas agreed to distribute leaflets to all pupils. The letters specifically asked "do you have asthma?" and actively sought volunteers for the study. This approach generated 68 asthmatics in 45 households, all of which were accepted into the study.

All house types have been subsequently identified and their technical details (materials, construction, orientation, site exposure, dimensions, heating systems etc) recorded and analysed using the NHER energy assessor software. This programme generates both SAP and NHER values which can be both norm and criterion referenced.

Fig. 5.1. Typical house-types used in the survey

HOUSE PERFORMANCE

To monitor the internal temperature and humidity fluctuations within each dwelling, thermo-hygrographic data-loggers (Tiny-Talk II) were deployed in all living rooms and bedrooms.Two temperature and relative humidity sensors were located externally at two addresses, in positions sheltered from direct exposure to the elements to act as micro-weather stations. The fuel consumption (electricity and/or gas) is also being recorded for each household. By comparing the fuel consumption, environmental profiles (internal and external) with the NHER and SAP ratings an estimation can be made to determine whether each house type is over/under performing in terms of energy use throughout the course of each year. Infra red thermography was also used to assess the integrity of the urea-formaldehyde cavity fill insulation, which is present in 70% of the dwellings in the study and was found in places to have significant voids. A sample of the housing types involved in the study is shown above.

The data-loggers were programmed on a 90 minute recording cycle which was geared to a 16 week downloading cycle. The initial data for October 1998 – January 1999 is shown below.

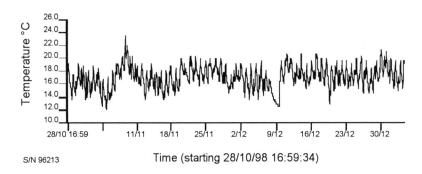

S/N 96213   Time (starting 28/10/98 16:59:34)

Fig. 5.2. Typical trace for indoor temperatures (coldest living room).

Results from the logging showed:
Coldest average living room temperature 17.0°C (min 12.0, max 23.4).
Warmest average living room temperature 24.5 °C (min 19.1, max 31.4).
Coldest average bedroom temperature 22.3 °C (min 10.6, max 18.1).
Warmest average bedroom temperature 22.3 °C (min 14.4, max 26.6).
Lowest average bedroom relative humidity 38.6% (min 21.5, max 57.3).
Highest average bedroom relative humidity 79.2% (min 64.1, max 93.7).

The initial monitoring cycles have confirmed high relative humidity levels in bedrooms. The initial NHER values confirmed that the dwellings are relatively energy efficient producing internal temperatures above the comfort threshold (18°C). When such relatively high indoor temperatures are combined with equally high internal water vapour pressures the research data produced to date would suggest that a high level of dust mite infestation would also be expected in carpets and bedding.

DUST MITE LEVELS

The logging interval selected dictated the timetable for the monitoring regime for all other data collection - up to 16 weeks. As dust mite levels vary with season [Spieksma 1967, Platts-Mills 1987], home visits are now being conducted on a quarterly basis to collect the following data:

- Downloading and restarting data loggers
- Dust collection
- Completion of a health questionnaire
- Collection of peak flow charts

    Using a conventional vacuum cleaner and a specially designed filter device, supplied by ALK (UK), dust samples can be readily collected from floors, beds, curtains etc. An area approximately $0.5m^2$ is vacuumed from the living room carpet, all asthmatics' bedroom carpets and all asthmatics' beds, for 1 minute. All dust samples are stored in the University Environmental Health departments freezer at -10°C immediately after each day of sampling to prevent any live mite activity occurring in the petri dishes.

    The vacuum cleaner bought for the site work is a simple tub vacuum (Goblin Aquavac – Type 7409P) with no internal filters. This was reasoned on the basis that any particles would be trapped in the petri dish attached to the filter device, rendering any internal filters redundant. Assay kits produced by Allert Biosystems limited were tried and tested prior to the initial home visit by the immunologists at Glasgow Western Infirmary. Each assay was performed twice initially to verify consistency of the assay kit and procedure.

    The dust samples are being processed at the Glasgow Western Infirmary's Immunology Department where immunological assays testing for Der p1 are performed. Results are returned in the form of µg Der p1 per gram of house dust. Results for the initial set of dust samples (October 1998 – start of 6 month baseline monitoring period) are shown below for the living room and bedroom carpets and beds of all participants.

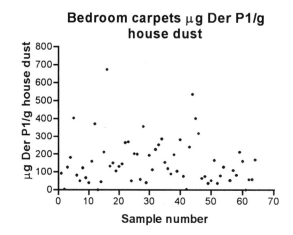

Fig. 5.3. Typical scatter diagram of dust-mite particle concentration in samples from bedrooms

The initial round of immunological assays confirmed that 87% of dust samples taken from the bedding were above the sensitisation threshold of 2μg Der Pl/g of house dust set by the WHO and 57% were above the upper threshold of 10 μg Der Pl/g of house dust. The figures for bedrooms and living room carpets were 73%/30% and 57%/20% respectively.

ASTHMA MONITORING

Peak flow meters were supplied to all participants who are required to chart and record the best of 3 morning and 3 evening peak flows - peak flow meters (see below) are simply a device for monitoring lung capacity – indicated on a scale on the side of a tube which the participants blow into (litres/minute). A simple numerical value is entered into a specialised table for each reading. All asthmatics are being asked to record major events in a life event diary. Events to be recorded include: changing of furniture (especially beds and sofas), illness, holidays, prescription changes, emotional distress etc. These diaries will be used as a reference to help explain any major deviations in peak flow readings throughout the study.

Finally, a questionnaire was adapted by the project team from the Euroqol [Dorman 1997] and MacMaster University questionnaires. As the present asthma research involves a large percentage of children it was felt necessary to omit many of the questions developed in the two existing questionnaires which dealt with purely adult topics.

5.9 Remedial strategies

The main thrust of this study is to identify whether the removal of dust mite colonies and their allergens from domestic dwellings has any effect on asthmatic symptoms.

STEAM

Various methods of achieving this goal were described earlier and the use of steam cleaning appealed to the team as it requires no expertise to operate the equipment. This is reasonably portable and incorporates the facility to kill mites and denature allergens (removing existing reservoir) within one integral unit. The system used was supplied by Medivac Healthcare Limited and is an industrial unit capable of producing steam at temperatures over 100°C and pressures up to 7 bar. Tests conducted by the manufacturer of the device claimed to have denatured between 80-90% of the total Der P1 sample. Trials conducted by the University of Strathclyde prior to application in participants' homes demonstrated a reduction in Der PI counts of between 79% and 99%.

As the highest mite levels are found in areas of high or constant occupation, all asthmatics' bedroom carpets, living room carpets and sofas were steam cleaned and then a set of dust samples taken after cleaning. Three rounds of assays were taken between October 1998 and April 1999, with a fourth set taken one week after steam cleaning had been applied in the two active groups. Results of the last assay will act as a measure of the efficacy of the steam cleaning to reduce Der PI levels below the known thresholds.

FANS

Mites require an absolute humidity greater than 7g/kg to survive due to the fact that most of their water is gained by osmosis [Hallas 1990]. They can survive short periods of time at lower absolute humidities, however if a level below the WHO value can be sustained over a long period of time, then no mites should survive.

The use of heat recovery fans has been incorporated into the project, with installation occurring in all participants' bedrooms. The fans themselves were supplied by BAXI, who were also involved in the Southampton asthma study conducted by Warner [Warner 1990]. The fans finally selected use a cross flow heat exchanger, which pre-warms fresh incoming air while stale inside air is simultaneously extracted.

Using this method, up to 80% of heat is recovered by the incoming air. As a result of this process pre-warmed dry air is supplied to the space and vitiated air is discarded from the room. Therefore, the ventilation unit effectively lowers and maintains the internal humidity reducing the risk of condensation dampness and the occurrence of mould. New technologies allow the fans to be almost silent when running at a continuous background level (in an average public sector bedroom this will produce an air change rate of 0.5 ach^{-1}) and thus the user compatibility may be greater than with the typical fan type currently being installed to simply provide wet zone extraction. A boost facility with an increased flow rate has also been provided and is user controlled.

BEDDING

Breathable mattress covers, new pillows and duvets have also been supplied to all dwellings, with the exception of the control group, to ensure a baseline of zero at the time of remedial strategy implementation.

SAMPLE SPLIT

The trial has been designed to be double blind and placebo controlled, and although the subjects are volunteers they have been randomly distributed into two active groups and a control, based on house type and tenure - consisting of the following measures:

Control group (CG): Placebo fans were installed in the bedrooms of all asthmatics. These fans have had the heat exchanger modified to ensure no cross flow of air, however the fans have been connected to the electricity supply and produce low level noise. Placebo steam cleaning was also undertaken.

Active group 1 (AG1): Placebo fans were installed in the bedrooms of all asthmatics. These fans have had the heat exchanger modified to ensure no cross flow of air, however the fans have been connected to the electricity supply and produce low level noise. Bedrooms and living rooms have been actively steam cleaned. New pillows and duvets were also supplied. AG1 will thus test the efficacy of the steam cleaning process by monitoring re-colonisation rates in furniture, bedding and carpets.

Active group 2 (AG2): Fans will be installed in the bedrooms of the asthmatics and connected to the electricity supply. Bedrooms and living rooms will be steam cleaned. New pillows and duvets will be supplied to all asthmatic participants. When the results from AG2 are compared with the other two groups, the efficacy of the fans can be demonstrated in controlling humidity and re-colonisation rates.

The placebo fans are placed on the external wall so that no external air is supplied to the internal space. Effectively, the fans are just recycling the internal air. The effect of these strategies will be evaluated on the following criteria:-

- Do any of the implemented strategies reduce and maintain this reduced humidity within the environment they are applied to?
- Is it possible to maintain humidity below the recommended level, for inhibiting the growth of dust mite colonies, of 7g/kg throughout the course of a year?
- If this threshold limit is attainable, does it have a detrimental effect on actual dust mite colonies within the domestic environment?
- If dust mite colonies are reduced and avoidance regimes applied to the domestic domain, is there a positive benefit in symptom relief for the asthma sufferer and for health in general, and if so what is the reaction time scale?
- What is the scale of the likely cost savings from such a preventative approach?

6

Impact of fuel poverty on health in Tower Hamlets

Lutfa Khanom

The multidisciplinary research described in this chapter has received joint funding from EAGA Charitable Trust, East London and the City Health Authority and the Local Authority (Housing Strategy Department). This study attempts to understand the relationship between fuel poverty and health and the impact energy efficiency interventions have on households suffering from fuel poverty and health problems. Detailed information is being gathered through a structured questionnaire and two in-depth interviews on people's perceptions and understandings of the relationship between their home heating practices, consumption of fuel and self-reported ill-health in the household. The research focuses on the response to and effects of energy efficiency advice in 24 households and capital investments in a few of the selected households in the public sector. The results are based on the data collected from the structured questionnaire with the baseline sample of 89, as well as extracts from the first stage of in-depth interviews with the final sample.

6.1 The sample

The sample was randomly selected on the basis that the respondents were claiming housing benefit. The sampling frame was the main database used within the Housing Department. 89 respondents out of a sample of 188 were interviewed within the areas of Bethnal Green, Wapping, Globe Town and Stepney Green. The final sample was selected from the 89 interviewed on the basis of the definition of fuel poverty that is provided by Brenda Boardman. She describes it as the 'inability to afford adequate warmth because of the energy inefficiency of the home' [Boardman,

1991]. She explained this as low income households not being able to afford to keep their homes adequately warm. Fuel bills are relatively high in comparison to the household's income, and/or they have reported problems with the living conditions of their home.

6.2 Basic links with health

Much of the recent research has looked at ill-heath and bad housing conditions within the public sector. In previous studies of fuel poverty, definitions of ill-health include social, physical and psychological manifestations. Hyndman's research on a survey undertaken in Tower Hamlets, East London, found relationships between damp housing conditions and subjective measures of ill-health. She found a high risk of dampness and condensation occurring in most cold homes (this is acute in partially heated homes). Hyndman argued that chest health is related to the humidity and temperature within the home [Hyndman, 1990, pp.131-141] Not only did she find evidence of sufferings of hypothermia, she also claimed there were reports of coronary, ceretoral thrombosis and respiratory diseases. Hyndman pointed out that dampness creates mould in the home, which carries a high risk of symptoms of undiagnosed 'hidden asthma'.

In addition to the impact on physiological well-being, there is also evidence to suggest effects on mental health. Arblaster and Hawtin [1993] have pointed out that the poorly designed buildings built in the 1960's have resulted in families now having to live in 'cold, damp, noisy and potentially dangerous property, often surrounded by rubbish and graffiti'. Most properties have little space for children to play, poor control over public stairwells and entrances, and their inhabitants have constant fear of vandalism and crime in the local area. Lowry [1991] states that psychological consequences are obvious when 'having to scrape mould off the walls of your house every day'. Cold, damp homes can cause emotional distress, increased anxiety, depression and irritability. With this in mind, this research does not examine causal relationships but focuses on the situation of households being in fuel poverty. Health beliefs and behaviours of households on low income and their coping strategies will be considered, in relation to fuel poverty.

6.3 The results

This section highlights some patterns from the data taken from the structured questionnaires and the first in-depth interviews. Several areas will be touched upon. These include: data on fuel expenditure; health problems; and correlations with the housing conditions in each household.

PAYMENT OF BILLS

Respondents were asked to provide details about their expenditure on fuel bills over a period of a year (broken down in quarters). They were asked how gas and electricity bills were paid and how much they pay. They were also asked whether they had difficulty in making ends meet.

Table 6.1. Percentage of income against percentage of fuel costs per week (random selection)

No. of people in Hhold	Estimated income per week (£)	Estimated fuel bill (gas + elec) pwk (based on 1/4ly bill n=x÷13)	Outgoing % of income on fuel per week
1	100.00	£16.96	16.9
2	100.00	£10.76	10.8
6	150.00	£15.38	10.25
2	100.00	£14.23	14.2
5	150.00	£15.38	10.2
7	200.00	£14.61	7.3
5	100.00	£12.69	12.6
9	200.00	£16.92	8.4
2	100.00	£14.46	14.4

Table 6.1 gives examples of income and expenditure from a random selection of the sample of 30. It has been suggested by Brenda Boardman [1987] that the cost of expenditure (per week) on fuel should not exceed 10% of the total household income [Burridge and Ormandy, 1993]. 7 out of 9 households above are spending more than 10%, 3 of which have an expenditure of 14% or more of their income on fuel. These households are smaller families (consisting of 2-4 members). This supports Neil Ritchie's [1990] view that the most affected tend to be elderly groups and single parents. Ritchie claims that these households heat their homes for 12-16 hours per day because they stay at home [Ritchie, in Hewett, 1996].

There are some figures which are below Boardman's suggested threshold, such as the family of 9 who spend 8.4% of their income on fuel bills per week. In this situation, the family claim that they tend to use and heat the living room more regularly than other rooms. The bedrooms are purely for 'night-use' (respondent No: 15). Most of these families find it very difficult to make ends meet. They tend to pay both the gas and electricity bills at one time, as well as other overhead costs such as rent, telephone bills etc.. Most families interviewed do not budget their income, primarily because they are not numerate and do not understand basic maths. As one respondent claimed:

'We do not budget what comes in the house and what is spent…it's too complicated…we spend money on whatever it needs to be spent on…this is how it works every week. When bills need to be paid, we struggle because we do not estimate how much money is needed for gas and electricity. At those times our priority is spending the rest of the money on food and other necessities…' (respondent 15).

HOUSING CONDITIONS

Table 6.2. Poor living conditions experienced by households (sample of 89)

Type of condition	No. of affected families (in each condition)	% in relation to total respondents interviewed
Mould on walls	28	31
Stained walls	39	44
Wallpaper peeling off	23	26
Damp carpets	13	15
Spoilt furniture	11	12
Spoilt clothes	11	12
Damp smell in house	41	47
Windows do not fit	18	20.5
Cold house	55	62

Please note that this table does not take into account the overlapping housing conditions experienced by each household.

To understand the overall situation of these respondents, it is important to look at the poor housing conditions experienced by the sample group. Table 6.2 highlights this situation. Out of 89 households, 23 respondents claim that their wallpaper peels off the walls, requiring them to change wallpaper on a regular basis, using money they cannot really afford to spend. Nearly half of these 23 respondents stated they had a damp smell throughout the house and that they felt the reasons for this were related to the poor conditions of the house. They also point out that the mould growth has caused regular 'damage to clothes and furniture' and that they 'replace items of clothing on a regular basis' (respondent no: 45), money which would otherwise have been saved or spent on other goods.

HEALTH PROBLEMS

Questions were asked in relation to the respondent's personal health as well as their children's health. The following two tables explain the findings. Attempts will be made to relate health problems to the housing conditions in the latter part of this section.

Table 6.3. Types of health problems by affected households over the past 12 months (sample of 89)

Health problems (respondents)	No. of affected families	No. of affected families (in % terms)
Blood pressure	15	17
Coughing	51	57
Aches and pains	41	47
Headache	56	63
Dizziness	18	20
Breathlessness	30	34
Depression	60	67

Table 6.3 sets out the types of health problems experienced by the respondents over the past year. Depression affects two-thirds of the total sample (67%). Respondents often said this is due to worrying about other severe illnesses such as diabetes, asthma or long-term illnesses, problems at home and with their children. A large number of respondents have regular headaches. 63% of the sample have stated that their headaches occur on a regular basis, but they are not aware of the reasons behind the regular pains. High blood pressure has occurred relatively more among the older groups (more women than men).

Table 6.4. Types of health problems associated with children of respondent over past 12 months

Health problems (children)	No. of affected families with children	% of affected families (as a % of 89)
Aches and pains	10	14
Diarrhoea	16	22
Wheezing	28	40
Sore throat	36	51
Vomiting	13	18
Irritability	22	31
Fever	48	68
Tantrums	16	28

Table 6.4 provides a brief understanding of the children's health problems through the eyes of the respondent. A large number of respondents have reported their children suffering from cold/flu-related symptoms. With fever affecting 68% of the total baseline sample and sore throat affecting 51% of the sample, the respondents said they are aware of some relation between the lack of warmth within the home. There is also a high number of children suffering from wheezing and skin irritability (40% and 31% respectively). Most parents argued that wheezing, coughing and other chest related illnesses tend to occur as a result of cold nights in the winter period. But it can also occur through breathing in the toxins created by mould and dustmites. This can also have an effect on skin irritability and most parents complained that the skin has been affected through touching the affected damp areas when children play or fight with their brothers and sisters. Parents have also said that some children miss school when such illnesses become severe.

22% of children have tantrums due to either the home being too cold or even too hot (when the central heating and gas fire are put on at the same time, the house can become hotter than that suitable for a child). Children from 13 families vomit on regular occasions (18%). These are new born babies to 5-6 years on average. Respondents were not aware of why their children are suffering from vomiting but do state that it is possibly due to the children not eating the right food and/or not eating meals at the right time.

6.4 Correlations between illnesses and housing problems (relating to aspects of fuel poverty)

The tables below outline the possible correlation between people's experience and perception of bad housing conditions and fuel poverty and their illness over the last 12 months. Households with and without housing problems were compared, looking at the number of reported health problems in each group. The tables outline the number of respondents who have encountered various problems in relation to the mould and damp conditions in the home. This is analysed with the reported health problems in general and/or related with the particular housing condition.

Both mental and physical illnesses have been taken into account. In addition, both the risk ratio and p-value have been inserted. This helps to assess the level of significance and confidence of the relationship between the two factors of self-reported ill health and housing conditions.

Table 6.5. Self- reported health illnesses (adult) with wallpaper peeling off

| | Housing condition | | | |
Reported health problem	Wallpaper peeling off	No wallpaper peeling off	Risk ratio 95% CI	P-value (X²)
Coughing	14/23	37/66	1.09 (0.73-1.61)	0.7
Depression	20/23	40/66	1.43 (1.12-1.84)	0.02**[1]

Depression was seen to be a long term issue after reports of many years of bad housing conditions. It is when depression and anxiety are compared to the long-term effects of wallpaper peeling off, that significance is found. Table 6.5 illustrates this relationship by explaining that 23 respondents reported to be experiencing wallpaper peeling off in their homes, and 20 from that group also claim to be depressed (RR=1.43, p<0.05).

Table 6.6. Self- reported health illnesses (adult) with cold home

| | Housing condition | | | |
Health problem	Cold house	No cold house	Risk ratio 95% CI	P-value (X²)
Coughing	35/55	16/34	1.35 (0.90-2.03)	0.1
Wheezing	21/55	3/34	4.33 (1.40-13.42)	0.0002***
Headaches	36/55	20/34	1.11 (0.79-1.56)	0.5

Table 6.7. Self- reported health illnesses (adult) with damp smell in home

| | Housing condition | | | |
Health problem	Damp smell	No damp smell	Risk ratio	P-value (X²)
Coughing	26/41	25/47	1.19 (0.84-1.70)	0.3
Wheezing	16/41	07/47	2.62 (1.20-5.71)	0.01*
Nausea	12/41	06/47	2.29 (0.95-5.56)	0.05*

Tables 6.6 and 6.7 have taken note of other self-reported illnesses such as aches and pains, wheezing, nausea and regular headaches in

[1] Please note for tables 6.6-6.9:　　*p = < 0.05
　　　** p = < 0.01
　　　*** p = < 0.001

relation to reporting having a cold home. In this situation, it is more likely that wheezing is a relating factor of a cold house (RR=4.33, p<0.05). Hyndman [1990] states that 'cold may affect the prevalence of viruses, influenza epidemics often occurring after a cold spell'.

Table 6.8. Self- reported health illnesses with overcrowded household

	Housing condition			
Health problem	Overcrowded household	No overcrowded household	Risk ratio	P-value (X^2)
Depression	21/25	39/64	1.38 (1.06-1.79)	0.03**

Overcrowding is another factor which has a significant relationship to depression. Table 6.8 shows that 21 out of 25 respondents claim to have both an overcrowded household and depression (RR=1.38, p<0.05). The presence of both wallpaper peeling off and living in overcrowded conditions can increase the relative risk of depression.

Householders who have reported a damp smell have a significant relation to other health illnesses. When examining the relationship between wheezing and damp smells in the home, a strong relation can be seen (RR=2.62, p<0.01). Dizzy spells are also another factor correlated with damp smells. There is an increased risk ratio of 2.29 (p< 0.05).

6.5 Children's health

Table 6.9. Self- reported health illnesses (for children) with mould on walls

	Housing condition			
Health problem	Mould on walls	No mould on wall	Risk ratio 95% CI	P-value (X^2)
Vomiting	8/25	05/45	2.88 (1.05-7.87)	0.03**
Wheezing	11/25	17/44	1.14 (0.64-2.03)	0.7
Coughing	9/25	11/45	1.47 (0.71-3.06)	0.3

Table 6.9 illustrates the common health problems of children which relate to other factors arising from fuel poverty. Wheezing and coughing are common symptoms recognised by health professionals when considering effects of fuel poverty. Ritchie argues that children living in damp houses, especially where fungal mould was present had higher rates of respiratory

symptoms, unrelated to smoking in the household. The rates of infection and stress were also high.

The most significant relationship is between children vomiting and mould on walls in the home. 8 out of 25 families with children suffer from vomiting (RR=2.88, p>0.03). It seems that the growth of mould on walls can increase the relative risk of children vomiting. Tenants Resource and Information Service (TRIS) supports this finding. In Action on Damp Homes, TRIS argues that moulds can trigger off allergies (similar to pollens triggering off hayfever during the summer) and can cause symptoms in sensitive people, such as sneezing, wheezing, vomiting and itching [TRIS, 1995]. They state that poisonous toxins are produced from the fungus created. This poison can contaminate food and can also be absorbed through lungs.

6.6 Conclusions

There are a number of weaknesses in the data presented which is
- from a small sample,
- relies on self-reporting of symptoms,
- lacks a control group and is
- taken only from the first stage of the research.

However the data does already highlight specific issues related to fuel poverty and its effect on respondents' health, as well as other issues.

1. Most respondents have difficulty in paying their fuel bills. Although the data illustrated weekly costs, most pay their bills quarterly, thus causing them to struggle at particular times over the year.
2. In addition many report that they have damp or cold-related problems in their house with and cannot afford to re-decorate their home. This does not help their physical or mental health.
3. As well as reporting common physical health problems associated with poor housing conditions among both adults and children, a large number of respondents have reported that they suffer from depression. This is also strongly related to poor housing conditions and respondents claim it is due to the 'helplessness' they feel at not being able to do anything to improve the living conditions for themselves and/or their children.

The respondents are heavily reliant upon the housing department but are reluctant to acknowledge the financial restrictions it faces.

The research is ongoing and results of the later stages, focussing on the response to and effects of energy efficiency advice and capital investments in selected households, will be reported in due course.

7

Tolerant building: the impact of energy efficiency measures on living conditions and health status

Geoff Green, Dave Ormandy, John Brazier, Jan Gilbertson

7.1 Introduction

There are apocryphal tales about the downside of comprehensive redevelopment of British towns and cities in the post-war years. But undoubtedly many municipal tenants of recently constructed high-rise apartments were advised by housing officials to turn up the heating, open the windows and cut down on washing if they wished to reduce the damp and cold conditions which were blighting their lives. This official line now has a tenuous scientific basis in the atmospheric balance model developed by the Building Research Establishment. In their report *Tackling Condensation*, the BRE describes how three key factors interact to determine 'home environmental conditions' - the weather, the building and the occupants [Garrat and Nowak, 1991, p.9]. So if the weather is an act of god and it is impossible to change the physical properties of a tower block – as officials in those early days claimed – then it follows that the occupants must be responsible for high levels of moisture production, low temperatures and poor ventilation, all combining to produce condensation. Furthermore, poor health resulting from such cold and damp conditions is in this sense, self-inflicted. Sonia Hunt condemns such casuistry as 'moral fragmentation occur(ring) when a collective problem is reduced to the characteristics of individuals.' [Hunt, 1993, p.70].

Towards the end of the 1960's boom in tower-block construction, high-rise tenants were put into a quandary. Naturally they sought warm and comfortable homes, appreciating how higher room temperatures could reduce condensation and damp conditions: but often they could not afford the extra fuel required. At first this fuel poverty was exacerbated by an increase in heating tariffs. UK energy costs soared when exporting states (OPEC) hiked the price of oil and the 1968 gas explosion in Ronan Point led to relatively expensive electricity becoming the dominant energy source in tower blocks. Later, as energy prices moderated, and households in general gained some relief, tenants of tower blocks continued to experience fuel poverty because of a relative decline in their incomes. The popular right-to-buy legislation introduced by a Conservative central government in 1980 had the effect over time of residualising the rump of unpopular municipal housing for occupation by low-income groups [Cole and Furbey, 1994]. High-rise 'social housing' in most cities became the preserve either of pensioners or of young people on state income support – with both groups generally unable to afford even the lower market price of adequate heating.

Initially, the campaigning response – often supported by municipalities - was to demand a further reduction in fuel tariffs for these vulnerable tenants so that they could increase their consumption and raise room temperatures. Later, a number of policy developments combined to change the perception of the problem - and switch the focus of attention - away from the lifestyle of tenants towards the energy efficiency of their accommodation. However, these policy changes led only to modest improvements in the UK housing stock [Boardman, 1991] until the 1992 Rio Earth Summit decisively changed the national agenda. Thereafter municipalities, effectively barred from new house building, developed technical expertise in retro-fitting their existing stock to help achieve national targets for reducing energy consumption. When making the case for capital investment, municipalities were also in effect claiming that tower blocks identified as 'unforgiving' or 'intolerant' of many domestic circumstances, could be turned into functional or 'tolerant' buildings. And there was a possible bonus that improvements in living conditions would lead to better health.

7.2 Aim of the Sheffield study

Prior to these developments, much of the empirical research effort 'had concentrated on the quest for explaining the nexus between inadequate housing and ill health' [Burridge and Ormandy, 1993, p.xii]. Academic enquiry tended to focus on the problem rather than the remedy. However, by the mid-1990's, there were great opportunities for evaluating whether *better* housing was indeed leading to *better* health. A number of large scale renovation programmes being undertaken by municipalities were systematically transforming the physical landscape of British cities and the living conditions of thousands of tower-block tenants. In a vulnerable district close to the centre, the City of Sheffield had pioneered a special combination of physical improvements in 4 tower blocks.

Our study focuses on this Sheffield experiment. Our overall aim was to establish the strength and significance of any links between living conditions and health status. Prior to our survey, the local residents' association were confident about the connection, believing that their hard work to achieve the right level and mix of improvements had delivered a significant health dividend for their members. Later, during the survey itself, our hearts were warmed by a letter from a 'highly delighted' resident who seemed to reflect the majority when he wrote 'All rooms nice and warm and free from damp and condensation. I am sure people will be better in health and spirit.' Figure 7.1 illustrates the pathways to health highlighted by the residents association and in the letter. Our task was to test their hypothesis.

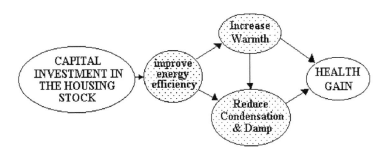

Fig 7.1. Pathways to health

Experts employed by the local authority had already calculated the increase in energy efficiency of an average apartment resulting from a capital investment of £29,000 per unit. Our investigation therefore concentrated on the other linkages. The study had three specific objectives - to measure:

1) the extent to which improvements in energy efficiency improve the warmth and comfort of residents;

2) the extent to which improvements in energy efficiency reduce damp and mouldy conditions;

3) the extent to which improvements in the living conditions - increased warmth and reduced damp and mould - are linked to better health status.

Certain complex relationships between the physical properties of the renovated buildings are beyond the scope of this study. Increased energy efficiency was generated by a combination of measures, with each having an independent effect on our two hypothesised determinants of health. About 22% of the cost was accounted for by replacing the underfloor electric heating system with a small gas-fired district central heating plant piping hot water to each apartment. These improvements alone would have encouraged tenants to raise room temperatures above the dew point of the indoor atmosphere [Markus, 1993] thereby reducing the risk of condensation.

About 40% of the cost was accounted for by better thermal insulation. Each tower block was encased in mineral wool insulation material with an outer skin of rainscreen cladding using an aluminium cassette-type system. This protects the old building structure from the weather and also prevents cold bridging caused by external exposure of concrete floors. Enclosing the open balconies with glass improved thermal efficiency still further. All of these measures tend to raise inside surface temperatures above dew point values so that condensation will not form on them. A final factor in the improvement package was a new ventilation system designed to replace vitiated air and remove moisture-laden air, while keeping heat loss to a minimum and avoiding draughts. For the purposes of this chapter, all these improvements are subsumed under the general heading 'energy efficiency' and combined into a single National Home Energy Rating.

7.3 Research design

Our evaluation was essentially retrospective. The tower blocks had already been refurbished when funding was secured for the study. The research design was therefore cross-sectional, with a survey of the improved apartments and their residents undertaken at single point after the intervention, rather than before and after. However, the design had two quasi-experimental features. First the key independent variable of energy efficiency had been calculated prior to the refurbishment and again afterwards. Second, a control was introduced by surveying a group of unimproved tower blocks with efficiency ratings comparable to the ex-ante position of the improved blocks.

In essence, the study would calculate warmth and comfort, damp and mould, health and quality of life in the improved compared with the unimproved tower blocks, then attribute differences to the investment in energy efficiency. Three survey instruments were used to elicit this information. Spot temperatures and moisture generating domestic activity were recorded in a diary/lifestyle account by residents over a two-week period. Damp and mould were assessed by building surveyors using a protocol derived from the English House Condition Survey. Health status was self-assessed by respondents using the SF-36 survey instrument embodied within a larger household questionnaire.

The SF-36 health survey originated in the USA [Ware and Sherbourne, 1992] but has been anglicised for use in the UK [Brazier *et al.*, 1992]. It measures health perceptions via 35 items measuring health across 8 dimensions, and one item measuring health change. People are asked for their own view of their health, and the questionnaire is usually self-completed, though it can be interviewer administered (as in this study). Responses to each item within a dimension are combined to generate a score from zero to 100, where 100 indicates 'good' health on each of the 8 dimensions.

The reliability of this instrument in terms of its internal consistency and stability over time has been tested and found to be satisfactory in meeting psychometric criteria [Stewart *et al.*, 1988; McHorney *et al.*, 1993 and 1994]. Proving its validity - in terms of demonstrating that it measures what it purports to measure - is more difficult, since there is no gold standard measure of health status. Nonetheless, there has been extensive testing in terms of the association between dimensions scores and type and severity of clinical conditions such as bronchitis, depression, back pain, peptic ulcers and other physical and mental problems, and other indicators of good health. The advantage of using this general measure is that it combines the consequences of a range of clinical conditions and their symptoms thought to be associated with housing.

A potential weakness of cross-sectional studies is the influence of confounding variables. Their presence makes it more difficult to attribute improvements in living conditions or health status solely to the specific refurbishment programme under review. Our study design therefore sought to isolate and then neutralise their effect by matching their parameters between the improved and unimproved properties and by matching the resident profiles.

Six potential confounders were controlled for in this way. *First,* climatic influences were neutralised (a) by selecting two groups of tower blocks less than 2 kilometres apart, in order to minimise geographic variation in external temperatures and (b) by undertaking all our surveys in the same winter period in order to discount seasonal variation. *Second,* in order to isolate the effect of a specific package of improvements, we chose a comparator group of tower blocks of similar age and construction. The first of the 4 one-bedroomed, 12 storey Hillside blocks - the improved group - was constructed in 1959. The last of the 3 two-bedroomed, 15 storey, Leverton Gardens tower blocks - the unimproved group - was completed in 1964. All were of reinforced concrete frame construction infilled with brick/block cavity panels. An electric heating system was embedded in the concrete floor of the living room and hallway and supplemented by a two bar electric fire.

By modern standards energy efficiency levels were low even when the properties were brand new. They were further eroded over the next 30 years as the fabric of the property deteriorated and the underfloor heating system malfunctioned or became relatively expensive to use. Prior to improvement, calculations of energy efficiency were undertaken for typical dwellings in both groups of blocks by Sheffield City Council's Department of Design and Building Services using both the National Home Energy Rating (NHER) and the rating using the UK Government's Standard Assessment Procedure (SAP). [Department of Environment, 1995]. On a scale of 0 to 10, the average NHER rating for an unimproved flat on *both* estates was 2.9 and the SAP rating was 28, compared with a local authority average of 34.4 in the 1991 English House Condition Survey [DoE, 1995]. After the improvements, the average NHER increased to 7.2 and the SAP to 68 (compared to between 80 and 85 for new buildings subject to current building regulations). Since there were no initial differences in average energy efficiency ratings for the two estates, we are confident that the great differences in energy efficiency ratings at the time of the study are attributable to the package of improvements implemented in the Hillside blocks.

The *third* and *fourth* controls concern domestic circumstances - two elements of lifestyle referred to in the introduction to this chapter. The schematic Figure 7.2 operationalizes the concept of a tolerant building. Our outcome measure of tolerance is the absence of damp conditions arising from condensation (distinguished from penetrating damp).

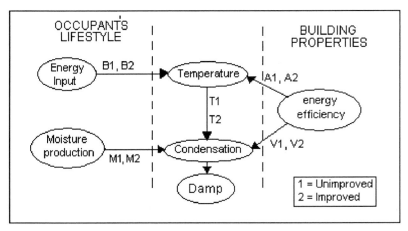

Fig 7.2. Controlling for lifestyle

Assuming similar weather conditions, condensation and damp in an apartment are essentially the product of the building's physical properties and the lifestyle of its occupants. A tolerant building will exhibit little or no damp irrespective of the lifestyle of its occupants – in other words, it will tolerate different levels of fuel consumption (energy inputs) and a range of domestic activity leading to different levels of moisture production. In order to test for tolerance, the research is designed to control for lifestyle and focus on the co-variation between the physical properties of the building (energy efficiency) and damp conditions.

Indoor temperatures are a function both of energy efficiency (A) and energy input (B). We expected average temperatures to increase significantly between period 1 prior to the improvements and period 2 after improvements. The study seeks to measure the increase in temperatures (T2 - T1) generated by the improvements in energy efficiency (A2 - A1). However, differences in temperature between the improved and unimproved flats may also be accounted for by differences in household consumption of fuel. This is likely to vary between households depending upon individual preferences and income constraints. But if there is no systematic difference between the two groups of tower blocks (B1 = B2), or if fuel consumption is actually lower in the improved blocks, then we may confidently attribute higher mean temperatures to the communal investment in energy efficiency.

Condensation and damp are a function both of energy efficiency - which includes ventilation systems (v) - and moisture production (M). A significant amount of moisture is generated by the everyday activities of residents and when energy efficiency is poor, this moisture may condense onto the walls and windows of an apartment to produce damp and mould conditions. However, our preliminary observations indicated that damp and mould had been virtually eradicated in the improved blocks but were significant in the unimproved blocks. The study was concerned to establish whether these differences could be accounted for by the more tolerant building properties generated by the improvement package. It was just possible of course, that the greater incidence of damp conditions in the unimproved blocks was generated by high mean levels of moisture production. If, however, the profiles of household moisture production were similar (M1=M2), then we may confidently attribute the eradication of damp conditions again to the communal investment in energy efficiency. Our formal hypothesis was that, compared with the unimproved properties, an improved mix of materials, design and heating elements in the improved property would interact with *similar* residents' lifestyle to produce warmer and drier conditions. In short, improved buildings can accommodate a diversity of lifestyles without ill-effect.

The *fifth* and *sixth* controls were introduced to neutralise the influence of socio-economic and demographic characteristics on differences in health status. Health inequalities persist in Sheffield as in other urban populations. Among its 29 electoral wards, standardised mortality ratios for the period 1981 to 1985 ranged from 78 to 117 and in 1993 from 71 to 114 (Sheffield + 100). These variations closely mirror the local geography of household income and deprivation as measured by the Townsend index. Long term illness is also related to class but also significantly to age; 58% of Sheffield residents over 75 reported limiting long term illness on Census day 1991 compared with the city population average of 16.1% [Central Policy Unit, 1993]. An early test of the SF-36 health status measure confirmed these gradients [Brazier *et al.*, 1992]; levels of perceived overall health in the city's population fall progressively through social classes 1 through V and increase progressively with age.

The study sought, therefore, to match the age and socio-economic profile of tenants in the unimproved properties with the profile of tenants in the improved properties under review. This was achieved not by stratifying the sample but rather by selecting tower blocks with similar settlement patterns. The early tenants of both the improved and unimproved blocks were a broad cross-section of working class tenants (Social classes III - V) displaced either by the slum clearance programme or encouraged into municipal housing by a decline in the private rented sector. Many, now retired (45.9% of the improved sample, 50% of the unimproved sample) remained as long term residents in the 1990's. However, this long established community was gradually being replaced by new and younger tenants who tended to be out of work. The history and dynamic of the populations of the two estates were therefore very similar. Three key control variables were almost identical in the two samples of residents surveyed. In both cases the age distribution was bi-modal with a mean of 56 years for the unimproved blocks and 57 years for the improved blocks. The mean household size was 1.2 and 1.1 respectively. The distribution of equivalised income - reported more fully elsewhere (Green and Gilbertson, 1999) - was almost identical with a household median of just over £101 a week, half the UK household income.

7.4 Results: Energy consumption and temperature

The headline result from our enquiry was that room temperatures in the improved tower blocks were substantially higher than in the unimproved tower blocks. Thermometers were placed in the main circulation area of half the flats on each estate (93 unimproved, 95 improved) and residents were asked to record the temperature at 19.00 hours each day. In order to control for climatic conditions, account was taken only of the residents who reported exclusively during the 21 day period between 12 February and 4 March 1995. Figure 7.3 shows the range of mean temperatures for individual flats - based on up to 14 readings. Profiles for the improved and unimproved flats barely overlap and the composite means (based on 1835 readings) are 22.1°C compared with 15.0°C - a difference of 7.1°C.

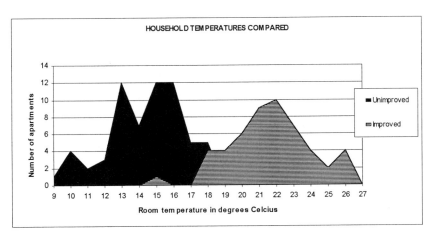

Fig. 7.3. Temperatures in the improved and unimproved blocks
Source: Household diaries. Improved n = 51 Unimproved n = 64

Lifestyle preferences and financial considerations will explain the variation in household temperatures *within* the improved and *within* the unimproved blocks. However the difference of 7.1°C in the composite means of each estate is probably the product of a systematic difference in energy efficiency. In order to discount the possibility that residents of the improved blocks are boosting temperatures by using more fuel than their unimproved counterparts, it is necessary to isolate and control for systematic differences in energy input into space heating. This can be difficult to achieve because households may use one fuel for a variety of appliances. However in our case study, the improved tower blocks presented a unique opportunity to distinguish different forms of energy use. The City Council's Department of Design and Building Services was already monitoring the use of the new district heating system, which provided just space heating and hot water. The only other energy source was electricity, which provided light and power to electrical equipment. For the whole of 1995, average consumption in each of the 184 apartments was 4238 kWh of useable energy. For the same blocks we derived an average of 2015 kWh of electricity consumed in those 108 apartments where reasonably accurate readings were supplied by the utility. When the two energy sources are combined they give an average annual energy consumption of 6248 kWh, as shown in Table 7.1. The table also estimates consumption for each quarter of 1995 (Q1 –Q4).

Table 7.1. Average energy consumption in an improved apartment

	Q1 kWh	Q2 kWh	Q3 kWh	Q4 kWh	annual kWh
Space heat	1316	442	86	1009	2858
Water heat	345	345	345	345	1380
Other electric	535	439	420	621	2015
TOTAL	2196	1226	851	1975	6248

The division of energy consumed by space heating and water heating was estimated by making three assumptions:
- in the summer month of July, space heating is zero and all the energy (115kWh) supplied by the district heating system was used for water heating;
- energy used for heating water was constant throughout the year at a rate equivalent to this July value (3x115 = 345kWh per quarter, 1380kWh annually);
- reflecting what residents told us, electric fires installed for supplementary heating were not used and energy for space heating was exclusively supplied by the district heating system.

The amount of energy consumed for space heating is calculated simply by subtracting the amount used for water heating (1380 kWh) from the total amount provided by the district heating system (4238 kWh) to give a residual value of 2858 kWh.

The single source of energy in the unimproved blocks was electricity supplied by a main meter and often also by an off-peak meter. The average electricity consumption was 6468 kWh of useable energy annually, for those 112 flats where reasonably accurate figures were supplied by the utility. A slightly more heroic assumption allows us to apportion consumption for space heating. Given the similar demographic profile and lifestyle of residents in the improved and unimproved blocks, they are likely to have similar patterns of energy consumption for both water heating and for powering electrical equipment. We have therefore simply taken these values for the improved blocks – for water heating 1380 kWh and for powering electrical equipment 2015 kWh – and imported them into the calculations for the unimproved blocks. Their sum of 3395 kWh annually is then deducted from total annual energy consumption of 6468 kWh to give a residual value for space heating of 3073 kWh. If the water heating system in the unimproved blocks is assumed to be 15% less efficient than in the improved blocks, requiring 2866 kWh to deliver the same amount of hot water, then the residual space heating calculation falls to 2886 kWh – almost exactly the same as in the improved blocks. These calculations are indicative only, serving only to make the point that significantly higher temperatures in the improved block energy cannot be attributed to higher energy consumption.

Table 7.2. Average energy consumption in an unimproved apartment.

	Q1 kWh	Q2 kWh	Q3 kWh	Q4 kWh	Annual kWh
Space heat	1405	507	101	1060	3073
Water heat	345	345	345	345	1380
Other electric	535	439	420	621	2015
TOTAL	2285	1291	866	2026	6468

7.5 Results: Moisture production, damp and mould

There were equally dramatic differences in the incidence of mould and damp conditions. A team of building surveyors from the School of Construction in Sheffield Hallam University independently assessed 126 improved and 137 unimproved flats using a modified form of the English

House Condition Survey. They calculated the percentage of damp and mould on each wall, ceiling and window frames. Only two of the 126 improved flats suffered a small amount of damp and this was confined to one room. In contrast over 40% of the unimproved flats had damp or mould conditions in one or more rooms. The headline figures are shown in Table 7.3.

Table 7.3. Unimproved apartments affected by mould or damp

	Percentage of apartments affected by damp in specified rooms				
	Living Room	Kitchen	Bathroom	Bedroom 1	Bedroom 2
Mould < 10%	23.4	22.6	21.2	21.9	24.1
Mould > 10%	6.6	10.9	7.3	12.4	12.4
Damp < 10%	27.0	23.4	23.4	23.4	24.8
Damp > 10%	5.8	9.5	6.6	13.1	7.3

Source: English House Condition Survey: N = 137. Note: Only mouldy or damp rooms are included, defined by any sign of mould or damp on either ceiling, wall or window frame; over 10% applies to any one or more of these surfaces.

Lifestyles and financial constraints will again explain some of the variation between apartments within the unimproved blocks. However, as with temperatures, the significant difference in the incidence of damp in the unimproved blocks compared with improved blocks is again probably the product of a systematic difference in energy efficiency. In order to discount the possibility that they can be accounted for by systematic differences in lifestyle, it is necessary to compare moisture production. This was undertaken via a lifestyle questionnaire. 111 households in the improved blocks and 94 in the unimproved blocks recorded their domestic activities of cooking, personal washing, dishwashing, clothes washing and clothes drying. The amount of moisture produced by each household was estimated from information in *British Standard Code of Practice for Control of Condensation in Buildings; BS5250:1989* and a Building Research Establishment Digest [BRE, 1985]. Fig. 7.4 compares the estimates for flats in the improved and unimproved tower blocks.

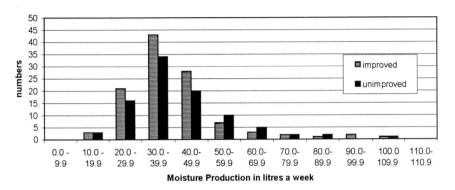

Fig. 7.4. Household moisture production
Source: Lifestyle Questionnaire. N = 205

The slightly larger average household size in the unimproved blocks accounts for marginally higher moisture production - a median of 38.2 litres and a mean of 41.5 litres a week compared with a median of 36.7 and a mean of 39.9 in the improved flats. Clearly these minor differences between the two profiles do not explain the large difference in the incidence of damp and mould. *Within* the unimproved tower blocks there is a small association between moisture production and damp conditions suggesting a building intolerant of certain lifestyles. Within the improved blocks there is obviously no such association, suggesting a building tolerant of a great diversity of resident lifestyles.

7.6 Results: Health status

Our study was designed to elicit any association between improved housing conditions and better health. There were two reasons for choosing a holistic measure of health as our principal instrument. First, the improvement package contained a number of different elements, each possibly contributing to a different aspect of health. Since we were not certain of how these elements would interact, we wanted a composite measure which would capture a range of possible health effects. Second, though there is accumulating evidence on the aetiology of specific diseases arising from cold and damp conditions (Strachan, 1993) the aetiology of better health arising from improved conditions takes us into largely uncharted territory. Without better evidence distinguishing the long from short-term disease effects of poor housing, we anticipated it would be difficult to capture the recovery from a specific disease arising from relatively recent housing improvements. We therefore took a pragmatic approach with our survey design, seeking general statistical association rather than a medical explanation.

Some of the problems associated with illness measures are evident from replies to the question on long standing illness imported from the General Household Survey. There was a significant difference (X^2: $p<0.05$, $N = 275$) between the improved blocks where 51.9% reported longstanding illness or disability, compared with 65.7% in the unimproved blocks. At first sight this appears to support an association between better housing and better health. However, some of this difference is accounted for by the question referring to *anyone* in the household with an illness. When only single person households were compared there was still a difference in those reporting long-term illness or disability – 48.4% in the improved compared with 56.5% in the unimproved blocks - but the difference was not significant at the 5% level. Even if it were, then it would still be difficult to infer that the level of longstanding illness in the improved blocks had *fallen* below the level in the unimproved blocks as a result of recent improvements. It is a moot point as to whether respondents in our Sheffield study or in the GHS equated the expression 'long-term' with 'permanent' and therefore 'irreversible.' If they did (and do) then subsequent improvements to the control group of unimproved properties would not lower the level of reported long-term illness or disability.

Our principal measure of health status was the SF-36, derived from answers to 36 questions embedded within our survey of individual residents. The origin of the measure, its reliability and its relation to clinical conditions is reported in the earlier section on research design. There were 275 responders and the percentage of missing data was extremely low, being between 0 % and 5.0%. For all 8 dimensions of the SF-36, the improved block had higher mean scores, indicating better health in comparison with the unimproved block - though General Health Perceptions were almost the same. Table 7.4 indicates that mean difference in physical role was significant at the 1% level and the difference in Energy/Vitality was significant at the 5% level. Differences in Emotional Role also bordered on significance at the 5% level.

Table 7.4. SF-36 Scores Compared

SF-36	Improved flats			Unimproved flats			p*
	n	mean	standard deviation	n	mean	standard deviation	
Physical Function	131	70.5	33.2	130	67.0	32.1	0.210
Physical Role	134	87.7	29.5	140	73.9	40.7	**0.003**
Emotional Role	135	90.1	28.8	140	84.5	33.6	0.055
Social Function	133	88.0	20.8	140	79.5	31.5	0.299
Mental Health	130	75.3	17.3	139	72.9	21.2	0.634
Energy/Vitality	133	59.9	21.9	139	51.9	26.6	**0.014**
Pain	128	77.3	26.7	138	70.5	30.4	0.077
General Health P	133	60.5	23.8	140	60.5	26.0	0.905

*p: Mann Whitney U Test Statistic p value N = 275

These results indicate that the SF-36 is appropriate to use on our study population. In contrast to other health measures, SF-36 has been able to find significant differences with a number of its dimensions. However these differences raise two related questions. Firstly, can they be attributed to housing improvements? Secondly, why are some dimensions significant and others not?

With a cross-sectional study design it is extremely difficult to establish attribution. In our earlier review of research design we stressed the requirement to match resident profiles between the unimproved blocks in order to control for confounding variables which might provide an alternative explanation of the difference in health status. In the event, the design successfully delivered two sample populations matched for age, sex, income and race. However, we were not able to match the proportions of residents in employment and therefore control in this way for the well-known 'healthy worker' effect. Significantly more residents (X^2: p< 0.05) of the improved blocks were in work (28.2% vs. 15.7%) than in the unimproved blocks and this might account for their better health. When those who work are excluded from the analysis, the residents of the improved blocks still score higher on 6 of the 8 dimensions of SF-36, and significant differences remain at the 5% level for the two dimensions of physical role and energy/vitality.

Secondly, assuming the hypothesis is true - that housing improvement accounts for differences in the score - then why are some dimensions significant and others not? Two issues have a bearing on the answer. Firstly, it might be hypothesised that certain kinds of housing improvements are more strongly linked to certain of the SF-36 dimensions than others. Secondly, better housing may take longer to make an impact on certain of the SF-36 dimensions than on others. In the event our evidence is not clear-cut. Besides better energy efficiency, the improvements included substantial security measures which made residents feel very safe in their homes. The impact of these measures on dimensions such as mental health, vitality and social functioning might be expected to occur more quickly after housing improvements. Our analysis shows there were indeed differences in social functioning and significant differences in energy/vitality, but relatively little difference in mental health. Most of the other housing improvements related to the energy efficiency of the property and might be expected to translate into an improvement in residents' physical condition. On the other hand, changes in physical functioning and pain are related more to physical morbidity and are likely to occur after longer delay, and indeed may take several years. In the event only a difference in Physical Role was significant.

One explanation of the small differences in mental health is that the elderly residents of both the improved and unimproved properties are a special band of 'survivors.' We have identified similar mental resilience in a similar group of elderly residents of tower blocks in Liverpool (Green, Gilbertson & Grimsley, forthcoming). In Sheffield they help lift the average mental health (MHI5) score of both improved and unimproved groups of residents above the average of a sample drawn from a cross-section of the city's population (Brazier, 1994). For most of other dimensions, the scores of residents in the improved group are above the City average and the scores of the residents in the unimproved group are below. Herein lies potentially our most important finding for the policy community. A group of residents with incomes way below the city average has a health status above the city average. And as far as we can tell at this stage, the only other significant positive difference between this group and the average city population is the excellent condition of their homes.

7.7 Conclusion

It is clear that capital investment in 4 Sheffield tower blocks has had a dramatic impact on the lives of their residents. A balanced package of improvements has transformed buildings which were unsupportive of low-income households and intolerant of typical lifestyles. As a result, the poorest household can now lead a comfortable life and any amount of washing, cleaning, cooking and drying can be undertaken without fear of creating damp and mould conditions. The liberating effect of a 'tolerant' building is evident not just from our objective data on temperature and building condition, but also from the subjective perceptions of residents themselves.

Downstream of these evident pathways to improved living conditions is the more problematic association with health status. On the face of it, the health of residents in the improved properties compares well with those in the unimproved properties and even with general population of Sheffield. These findings could have enormous significance for public policy, for they indicate that housing investment may break the vicious circle of low incomes, poor housing and poor health. However, we have qualified our results because of the comparative design of the study. Despite controlling for income and age we do not have the necessary degree of confidence to attribute elevated health status to improved living conditions. For this reason, the team has embarked on a more powerful longitudinal study of housing investment and health in Liverpool.

We are more confident about the general validity of our evidence connecting the upstream variables of housing investment and improved living conditions. Our study supports the findings of the 1991 English House Condition Survey, which indicated that residents respond to improvements in energy efficiency by increasing warmth and comfort rather than reducing energy consumption. Not all improvement packages will be as well balanced as in Sheffield. There is evidence from elsewhere that different improvement packages have not delivered cost-effective results. In cities where neglect has led to structural weakness then the cost of sustainable improvements may be prohibitive. However, in our view many tower blocks in the UK and across continental Europe are suitable cases for treatment. If there is demand from future generations, then the right improvement package – shaped by residents as well as professionals – can be cost-effective in transforming dysfunctional buildings into comfortable homes.

Part Two:

Tools for research and practice

8
The Affordable Warmth Index

9
Mould Index

10
Winter morbidity and fuel poverty: mapping the connection

11
Modelling the health cost of cold damp housing

8

The Affordable Warmth Index

Jake Chapman and Brian Scannell

The Affordable Warmth Index (AWI) provides a simple yet accurate assessment of whether a householder can afford the energy required for their specific property. The assessment is carried out using a computer program based on technology developed for calculating Energy Ratings, which are already widely used by social landlords and fuel poverty groups. The assessment software can also be used to explore the energy efficiency improvements required to reduce the energy requirements of the householder to a level they can afford. The aim of the AWI is to provide a quantitative tool to all those involved in tackling fuel poverty so that the problem is better understood and the best solutions and policies developed, in order that resources are effectively targeted.

This chapter describes the background to the AWI and the current technical specification. Detailed field trials are currently underway, therefore some changes may occur when the final version is released for general use in late 1999.

8.1 Introduction

The National Home Energy Rating (NHER) scheme is a professional energy rating scheme, enabling member organisations to qualify individuals to use software to assess the energy efficiency of a dwelling. The core membership of the NHER scheme includes over 200 Local Authorities, over 140 Housing Associations, over 960 builders as well as fuel utilities and energy consultants. The original idea for an Affordable Warmth Index arose at a conference discussing ways in which the NHER scheme could be extended so as to be more useful to its members. The development of an index for quantifying fuel poverty was the most important way in which this group wished to see the energy rating system extended.

There are two commonly used energy ratings, both of which provide a measure of the energy efficiency of a dwelling. The oldest and most comprehensive is the NHER itself which is based upon the total annual running costs per square metre of dwelling. There is also an official government rating, the Standard Assessment Procedure (SAP), which is based solely on the space and water heating costs per square metre and is independent of location.

The core technology that underlies energy ratings is the model used for calculating the energy requirements of the dwelling. This is the Building Research Establishment Domestic Energy Model, BREDEM [Anderson *et al..*, 1985]. There have been many different versions of BREDEM, though only three core types are now in use. The first is BREDEM-12 which is the Building Research Establishment's preferred method for calculating annual running costs. It is used in the NHER scheme [Chapman, 1991] and has been developed to ensure accurate assessment of total running costs [Anderson *et al.*, 1996]. The second is BREDEM-9 which is a simplified version of BREDEM-12 used in the SAP calculation. The main simplifications are the exclusion of energy use for cooking, lights and appliances and consideration of location. Other factors are excluded because they were judged too complex for the SAP worksheet - for example the assessment of the effect of having stairs opening directly into the main lounge. The third version of BREDEM in use is BREDEM-8, which is a monthly model and is better able to handle situations in which there are high solar gains or other strong seasonal fluctuations.

All the versions of BREDEM have been subject to detailed testing against field data and shown to be very reliable. In particular BREDEM-12 and BREDEM-8 have been shown [Henderson *et al.*, 1986, and Dickson *et al.*, 1996] to predict running costs to within about 10% in well heated dwellings.

Energy ratings and the associated technology have been widely accepted by building professionals and have facilitated significant improvements in standards of refurbishment and new-build in the social housing sector. Energy ratings have also become standard fare amongst builders since the introduction of the SAP rating into Part L of the 1995 Building Regulations. They are even poised to impact private householders with proposals which would require mortgage lenders to provide ratings and advice with all valuation surveys; indeed some lenders already provide this service. But overall they have had much less impact on fuel poverty and the very poorest housing.

One of the reasons for this lack of impact is that energy ratings were deliberately designed to be independent of the way that occupants used the dwelling; the rating was a measure of the energy efficiency of the dwelling fabric and the installed heating equipment. Associated with this is the assumption built into the energy rating evaluation of a "standard occupancy pattern"; a prescribed pattern of heating the dwelling, use of hot water and appliances and a standard number of people for a given dwelling floor area. This is akin to the standard driving cycle used to assess mpg figures for cars. Although the standard usage pattern reflects the average behaviour in well heated properties it does not reflect at all the issues and problems where occupants cannot afford adequate heating. What is more, the estimates of consumption and savings available made using the standard usage pattern would not apply to low income households living in dwellings with poor insulation and heating equipment. This is the area that the AWI set out to address.

8.2 The development of the AWI

During 1998 an Advisory Group was established to develop the AWI in detail. The group included representatives from NEA, Local Authorities, Housing Associations, academia, energy consultants and the fuel poverty lobby. The Group addressed a wide range of technical and practical questions. Amongst these were the following:
- Should the index be related just to heating bills or to total fuel bills?
- What assumptions should be made about the way that low income households heat a property?
- Should the index relate to an absolute cost (in £/week) or as a percentage of income?
- Should the system take into account the difficulties associated with housing small families in properties too large for their needs (and therefore very expensive to heat)?
- Who would use this index and in what ways?
- Would the quantification of the problem really help the fuel poor?
- Would rent officers with targets to meet use any such system and people desperately wanting any accommodation, no matter how affordable it was to heat?
- Could the AWI be used to increase the effectiveness of HEES or discretionary grants?

In answering these and related issues, what emerged was an outline technical specification for the AWI which was commented on by the Advisory Group prior to being implemented in software. After further comments a version of the AWI was agreed late 1998, and is currently employed in ten field trials throughout the UK. The aim of the trials is to provide detailed feedback from actual users on both technical issues, such as the data requirements, and the effectiveness of the AWI in practice. These applications, together with preliminary results from one of the studies, are described in section 4 below. Since early 1999 the development of the AWI has been supported by grants from the EAGA Charitable Trust and EAGA Services Ltd.

8.3 The technical definition of the AWI

The most widely accepted definition of fuel poverty is the need to spend more than 10% of income on fuel. Unfortunately this definition has tended to be used without clearly defining "income", "need" or "fuel".

The Advisory Group concluded that the Affordable Warmth Index should express the link between household disposable income and total fuel running costs based on actual occupants and standard comfort and non-heating energy use levels. It also concluded that the index should be expressed in a way which was both easily understood and provided the necessary sensitivity to ensure it was useful.   Each of these issues is discussed further in the following sections.

INCOME

The AWI is calculated on the basis of the householders' disposable income, exclusive of housing costs. As part of the development, standard tenant groups have been identified and the assumed income level for each group of occupants is based on the benefit payments that can be claimed, excluding housing allowances. Housing allowances are excluded on the grounds that they are not part of a disposable income. This exclusion means that when the AWI is used for householders who are in employment, their housing costs have to be deducted from their income.

The reason for excluding housing costs is to avoid overstating income levels for householders living in areas where housing costs are high, in particular urban metropolitan areas. A householder in such an area may be receiving a high level of housing allowance, however since this is paid directly to the landlord in most instances, it does not represent a controllable expense on the part of the tenants.

Including housing allowances could result in householders on minimum benefit, living in poor housing, being excluded. For example, consider two householders both on minimum benefit level and living in similar properties but with one in a high cost area and the other in a low cost area. Including the housing allowance as part of their income could result in the one in a low cost area being identified as in fuel poverty, whilst the one in a high cost area is not. Clearly this is not acceptable.

The minimum benefit levels can be established from the published information since for each category of benefit there is a defined level of total benefit. The current values implemented in the software are set out in Table 8.1 below. Note that in the software, these data are stored in a separate file, so they can be updated without changing the programs themselves. It is envisaged that once the system is finalised the relevant data file will be available for downloading from an Internet site, thus allowing users to update as often as necessary.

Table 8.1. Current values of benefit used in the software

Occupant group	Benefit (£/week)
Single person	38.90
Single parent with one child	81.80
Single parent with two children	98.70
Couple with one child	104.85
Couple with two children	121.75
Single pensioner	68.70
Pensioner couple	106.80

FUEL RUNNING COST

It was decided to base the Index on *total* running costs rather than just space and water heating costs because:
- For the low income groups it is the comparison of the total fuel bills with the disposable income that matters. There is no sense in comparing part of the fuel bill with total income.
- Most of the non-heating uses of fuel are non-discretionary, especially lighting, cooking and appliances such as refrigerators, kettles and TV. This means that, in practice, low income groups will control their total expenditure by reducing the space and water heating costs.

The total fuel running costs for a property depend upon:
- The external fabric of the dwelling, particularly U-values and air-tightness.

- The fuel used and system efficiencies for providing space and water heating, including the effects of controls on these systems.
- The location and exposure of the dwelling (determining the external environment).
- The extent, duration and level of heating and the level of use of hot water, cooking and other appliances by the occupants.

Items 1 to 3 are all directly related to the physical characteristics of the property and the installed heating system, whereas each of the elements in item 4 is dependent on the occupants and their behaviour. In the BREDEM-12 model, these are referred to as occupancy factors and under normal circumstances, the model uses a set of "standard occupancy" assumptions to predict average fuel running costs. Hence for a specific type of house, the use of the standard occupancy assumptions ensures an accurate prediction of the average fuel running costs for that type of dwelling. However, observations of actual fuel running costs for an estate of identical houses of this type would show a five to one variation in actual fuel use, reflecting the effect of variations in actual occupancy factors compared with the standard occupancy assumptions.

Predicting the energy use in low income households is particularly problematic since:

- These groups tend to have more erratic heating patterns making it difficult to define crucial variables such as the demand temperature (i.e. the temperature to which the house is heated).
- These groups also tend to heat different parts of the house at different times, they rarely heat the whole house together. This means that the internal heat flows between rooms can become as significant as the heat losses to the environment in determining temperatures and energy demands.
- the properties are often in poor repair and hence will not conform to assumptions about average ventilation rates or standard U-values.

There is not much published information on actual occupancy factors in low income households, therefore suitable occupancy factors have had to be defined for each of the potential occupant groups. It was agreed that these occupancy factors should be based on desired comfort levels for heating, hot water, lighting, cooking and appliance use. The general principles on which the occupancy factors have been developed are as follows:

- The heating *temperatures* are assumed to be the average values for well heated properties. It is known that the fuel poor heat to lower temperatures, but the definition of affordability should be based upon the average enjoyed by the rest of the population.

- The heating *periods* should reflect the actual lifestyles of the people involved. Thus it is to be expected that people with young children, the elderly and the infirm will require heating all day.
- At present the *cooking and hot water use* is kept the same as the average use by a family of the same size as the defined tenant group.
- The *electricity use* is defined for each group separately.

The demand temperatures and hours of heating, have been estimated by examining a number of sources, but is recognised as an area which requires further research.

On the basis of available information, it was considered that there were no grounds for changing the demand temperatures from those used in the BREDEM standard occupancy factors. These specify a demand temperature in the main living area of 21C and 18C in the rest of the property. As mentioned previously, available information suggests that low income groups do not generally heat their properties in this way. Frequently the demand temperature, especially outside the main living area, will be lower and a complete absence of heating outside of the main living area is relatively common. However it was strongly felt that the AWI should not perpetuate the concept of differential standards for low income groups, therefore the assessment of affordability assumes the same desired levels of thermal comfort as is recognised for the population at large. Indeed, consideration was given to specifying a higher demand temperature for occupant groups containing an OAP, but it was agreed that there was insufficient evidence to justify such a decision at this time.

The main source used for the hours of occupation for the property was "Activity Levels within the Home" by Brenda Boardman. This is a 1985 paper but is based upon actual survey information and distinguishes between different groups. Further information on occupancy levels may be available from the English House Condition Survey (EHCS). The latest published survey is for 1991, the 1996 report is due out soon and data has been requested from this source.

After examining the available data it was felt that there was not enough evidence to warrant assigning a different number of hours of occupancy to each occupant group. Instead a few standard patterns have been defined, for example one pattern represents being in all day, and then using the best pattern for each group. This avoids creating an additional source of variation in the definition of the AWI and makes the results easier to interpret. Note that within the software implementation of the AWI it is possible for the user to specify all these occupancy variables manually if required.

The 'all-day' heating pattern assumes heating from 8am to noon and from 3pm to 11pm each weekday; at weekends the heating is assumed on from 8am to midnight continuously. The single person heating pattern is 8am to 10am and 6pm to 11pm each weekday and 8am to noon and 5pm to midnight at weekends. These represent average levels of occupancy and are consistent with the data currently available.

It was agreed that there are currently no grounds for changing the standard method used by BREDEM-12 to calculate hot water or cooking fuel use for any of the occupant groups. The standard method is based on the number of occupants. In practice, the available historical data suggests that low income households typically use less fuel for these purposes than higher income groups. However, as argued in the case of the heating temperature, the point of the AWI is to enable policies to be developed that enable these groups to be able to afford to enjoy the generally accepted standards of comfort and amenity associated with fuel use, rather than perpetuating under-use.

Finally, the calculation of the electricity use by each tenant group is very significant since:

- Most electricity use is non-discretionary.
- The electricity bill is likely to be between 30% and 60% of the total fuel bill paid by the occupants.
- The "standard occupancy" assumptions about electricity use make unrealistic assumptions about levels of appliance ownership for these tenant groups. There is also an extremely large variation in electricity use within identical household groups, making its accurate assessment difficult.

The procedure adopted has been to make use of fixed levels of electricity use for each tenant group, irrespective of the size of dwelling in which they are living. This is in contrast to the standard BREDEM-12 equations, which predict electricity use as a function of both the number of occupants and the dwelling floor area. The electricity use in lights and appliances is taken as 800 kWh for a single person, 1200 kWh for a single parent with one child and 2500 kWh for a couple with two children. As with other aspects of occupancy the user can specify a different level if required.

AFFORDABLE WARMTH INDEX SCALE

The AWI numerical scale value was originally calculated using the following formula

AWI = weekly income × 52 × 10 / Total annual fuel bill

This scale ensured that the AWI equals 100 when the Annual Fuel Bill represents 10% of the total disposable income. Using this formulation the scale value can never become negative, which is regarded as a benefit since negative values for an Index are difficult to interpret. In common with the NHER and SAP scales a higher index value is better.

The relation of this AWI to both income and percentage of income is illustrated in Figures 8.1 and 8.2 below. Unfortunately, this scale has poor discrimination in the important areas of fuel poverty, namely when the fuel costs are in the range 15% to 30% of disposable income. For example it can be seen that changing the percentage of income spent on fuel from 15% to 30% changes the AWI from 66 to 33 - a change of 33 units. Changing the percentage from 15% to 7.5% changes the AWI from 66 to 133 - a change of 66 units. Thus the scale is most discriminating in the less interesting area, namely where fuel poverty is less severe.

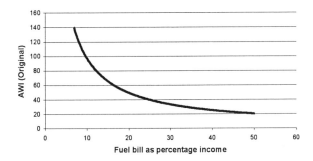

Fig. 8.1. Original AWI versus percentage income

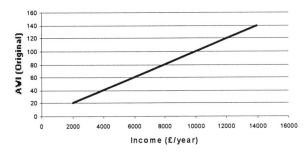

Fig 8.2. Original AWI plotted against household income (based on annual fuel bill of £1,000).

A range of different functional relationships were explored in order to derive a scale that was more discriminating in the area of interest, that had a strong relationship to the Government banding of fuel poverty and which gave an AWI of 100 when fuel bills were 10% of disposable income. The best solution to this specification so far found, and currently incorporated into the AWI being piloted, is

$$AWI = 140 - 4 \times P$$

where P = *fuel costs as a percentage of income*
        = *(total fuel costs × 100) / (52 × weekly disposable income)*

This formulation is subject to the condition that the AWI is set to zero for all cases where the fuel costs exceeds 35% of income. The relationships are illustrated in Figures 8.3 and 8.4.

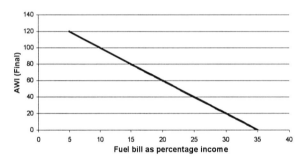

Fig 8.3. Final AWI versus percentage income

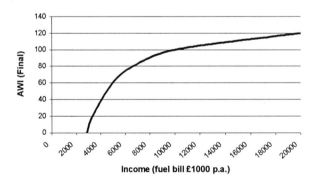

Fig. 8.4. Final AWI versus household income (based on an annual fuel bill of £1,000)

The government has introduced a categorisation of fuel poverty based upon the percentage of income spent on fuel, although there is still some debate about whether total or disposable income should be used and whether all fuel uses should be included. For the moment it has been assumed that following consultation, the government definition will follow the conventions agreed in developing the AWI. The relationship between the government defined categories and the AWI is set out in the table below. Note that this definition of the AWI provides adequate discrimination in the regions defined by the government categories.

Table 8.2. Relationship between government defined categories and AWI

Income needed for comfort	Category	AWI value
Less than 10% of income	Affordable	> 100
10% to 15% of income	Marginal Fuel Poverty	80 to 100
15% to 20% of income	Fuel Poverty	60 to 80
20% to 30% of income	Severe Fuel Poverty	20 to 60
Over 30% of income	Extreme Fuel Poverty	20 <

8.4 Applications of the AWI

The case studies currently underway to demonstrate the utility of the AWI usefully reflect the range of application that exists for the Index. The following section describes the one case study for which results are available. A summary of the other case studies is included in an Appendix.

PENSIONERS' ENERGY PLAN

Since 1996, NES has been monitoring the Energy Saving Trust's Pensioners' Energy Plan project (PEP). PEP is managed locally by Home Improvement Agencies and Local Authorities, and is aimed at pensioners who own their homes, but have little capital savings available for home improvements. PEP helps these pensioners to contribute to the cost of energy efficiency improvements by releasing the capital locked up in their homes. Other home improvement works are often carried out at the same time, for which there is often grant assistance available. In the pilot of the PEP programme, 66 detailed data sets were collected before and after work was completed. EST has agreed that NES can use this data for a case study that will analyse this data in light of the AWI. The initial results from this case are set out below.

Figure 8.5 indicates the improvement in the AWI realised by the project, with an improvement in the average AWI from 43 to 66. There is a comparable improvement in the SAP rating; the average before improvements was 30 and after all the measures this increased to 49. However, as shown in Figure 8.6, there is no significant correlation between the SAP and the AWI (the figures shown are for the properties prior to the improvement measures). This is to be expected since the SAP is independent of both occupant group and floor area, both of which have a substantial effect on the AWI. Whilst the SAP is not an effective surrogate for the AWI *per se*, there is a significant correlation between the improvement in SAP and the improvement in AWI, as illustrated in Figure 8.7.

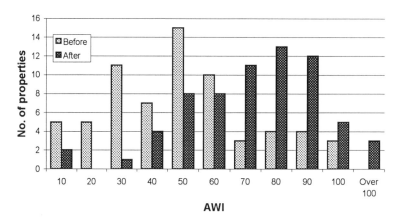

Fig. 8.5. Improvement in AWI

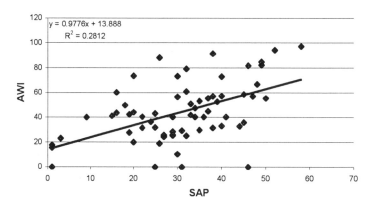

Fig. 8.6. AWI versus SAP before improvements

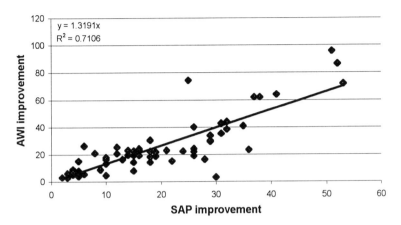

Fig. 8.7. AWI change versus SAP change

8.5 Conclusions

On their own energy ratings have not saved any energy. However by introducing a standard method of assessment and providing reliable and accurate estimates of running costs and savings, energy rating technology has transformed the energy efficiency industry. One very visible result is that there is now a language used by professionals for comparing the energy efficiency of houses.

It appears that the AWI may provide timely support for the government's initiative to reduce fuel poverty in a similar fashion. It provides a reliable and soundly based quantitative assessment of fuel poverty and as such gives both a language for comparison and tools for assessing the most efficient means of reducing fuel poverty. The results from the case studies are likely to show areas, such as discretionary grants and the design of refurbishments, where the software can assist directly in developing effective fuel poverty policies.

Appendix 8.1: AWI case studies

EAGA
As well as co-funding the project, EAGA is also participating in a case study. It has been agreed that HEES installers will be asked to collect a limited amount of additional information relating to occupancy for the households where they install measures. This information will then be used to see what effect the work carried out had on affordable warmth. There is also scope to look at "what if" scenarios, such as if more measures

were installed or different measures were available. This is in-line with the current review of HEES, and will highlight the benefits and disadvantages of allowing those in fuel poverty to have more than one HEES package installed, for example.

NATIONAL ENERGY ACTION (NEA)

Since July 1996, NEA have been operating a project in North Benwell, Newcastle upon Tyne, which aims to improve the energy efficiency of low income households using an energy audit based approach to determine the most cost-effective measures to install. It is part of a larger regeneration initiative in an area of early twentieth century housing, and provides direct benefits to individual households. Under the case study, the NHER data, along with further information collected regarding occupancy, will be entered into the AWI software, to investigate whether different home improvement recommendations would be made if the affordable warmth of the property were taken into consideration.

CAMBRIDGE HOUSING SOCIETY (CHS)

Cambridge Housing Society charges rents on all their self-contained homes in accordance with an attribute based rent policy. An attribute already exists for NHER ratings. This policy broadly aims to set rents which reflect the quality of the property and the differing average incomes for tenants. Rent levels are restricted to approximately 25% of tenants' disposable income, and CHS are keen to investigate forging a similar link between running costs and tenants' disposable income. By linking the AWI to rent values, it would be possible to carry out works to improve the energy rating to a house and charge an increased rent of perhaps half the anticipated savings in running costs

CAMBRIDGE CITY COUNCIL

Cambridge City Council has recently carried out energy audits on 41 properties that have applied for Grant Assistance under the Housing Grants, Construction and Regeneration Act 1996. During the case study, the Council will review these and any further applications made in the next two months, in light of the Affordable Warmth Index, to ascertain how Grant assistance can be targeted as a priority to those homes shown to have heating which is unaffordable.

MID SUFFOLK DISTRICT COUNCIL (MSDC)

An ongoing energy improvement programme is being undertaken by Mid Suffolk District Council, a rural local authority. The case study will examine how the Affordable Warmth Index could be used to better focus and advance the improvement programme. In the private sector, the Council will look at how the existing grant system could be modified to give grants to upgrade properties to improve their affordability for the residents. In MSDC's own stock, refurbishment schemes will be considered to see how cost effective measures could be incorporated to make the improved properties more affordable for the tenants.

ABERDEEN CITY COUNCIL

Aberdeen City Council have recently launched a Secured Loans Scheme targeted at owner occupiers who are on low, fixed incomes, have no access to capital and are experiencing fuel poverty. A pool of capital has been established by the Council and administered by Care and Repair, to fund energy efficiency measures and central heating installation in such properties. The loans are repaid when the house is sold. In association with SCARF, Aberdeen City Council will use the AWI to help identify and prioritise potential clients for this Scheme.

DONCASTER METROPOLITAN BOROUGH COUNCIL

Doncaster Metropolitan Borough Council are currently running a "Healthy Homes" scheme with the Health Authority, through which some private sector grant assistance has been ring fenced in order to target, with a health emphasis, those in fuel poverty. The Council takes referrals from Health professionals, and then sends a surveyor to the property to carry out an energy audit. Recommendations are then made for energy efficiency improvements, and a schedule of work is drawn up. Through the case study, comparisons will be drawn between the advice currently given to clients, and the advice that would be given if the AWI were used on a regular basis. The case study will also investigate the differences that would be made to the property and the tenants if a different package of improvements were recommended, taking the fuel poverty of the occupants into account.

WARM FRONT PROJECT, ECSC

ECSC is currently operating the Warm Front project in the Brighton and Hove area, providing training and support to voluntary sector networks in order to deliver advice and stimulate grant uptake in the domestic private rented and owner occupied sector. The results of the surveys will be entered into the AWI software and considered with a view to examining how proposals for targeting and modifying criteria for grant assistance in this sector could be developed.

FIRST REPORT

First Report is a consultancy company specialising in energy audits and research. They have already undertaken extensive energy analysis on a block of flats in Liverpool, occupied predominantly by pensioners, one of the social groups most at risk from fuel poverty. The research was undertaken with Sheffield Hallam University and also examined the relationship between cold homes and ill health. The data collected will now be entered into the AWI software, and analysed with a view to recommending the best set of energy efficiency improvement measures for the tenant group.

9

Mould Index

Tadj Oreszczyn and Stephen Pretlove

"then he who owns the house shall come and tell the priest, 'There seems to me to be some sort of disease in my house.' Then the priest shall command that they empty the house before the priest goes to examine the disease, least all that is in the house be declared unclean; and afterward the priest shall go in to see the house. And he shall examine the disease; and if the disease is in the walls of the house with greenish or reddish spots, and if it appears to be deeper than the surface, then the priest shall go out of the house to the door of the house, and shut up the house seven days. And the priest shall come again on the seventh day, and look; and if the disease has spread in the walls of the house, then the priest shall command that they take out the stones in which is the disease and throw them into an unclean place outside the city; and he shall cause the inside of the house to be scraped round about, and the plaster that they scrape off they shall pour into an unclean place outside the city; then they shall take other stones and put them in the place of those stones, and he shall take other plaster and plaster the house."

LEVITICUS 14 35-43

Mould is not a new phenomenon in buildings as the above quote from the Bible indicates. One of the main health issues often related to affordable warmth is that of mould growth and dampness in dwellings. This chapter reviews the current results of research into the causes of mould growth, proposes a mould index for dwellings and investigates the impact of fuel poverty on the occurrence of mould growth.

9.1 The problem

Approximately 20% of all dwellings in England suffer from mould growth and dampness to some degree. This number of incidences of mould growth and dampness in private rented dwellings, so often occupied by the fuel poor, is significantly higher than in any other sector [DOE 1996].

There have been many studies that suggest a link between health and dampness and mould [IEA 1991]. In fact one could argue that the causal link is perhaps stronger than between low internal temperatures or fuel poverty and health. However, it is always difficult to identify a causal relationship because of the many confounding factors.

Mould growth is not the only potential health hazard resulting from dampness and high internal relative humidities. The House-Dust Mite (HDM) which thrives in high relative humidities is considered to be an important causal agent not only for asthma, but also for other allergic disease, such as atopic dermatitis, rhinitis and keratoconjunctivitis [Cunningham 1996]. The MRC Institute for Environment and Health conclude that *"There is clear evidence that antigen derived largely from mite faeces is one of the major causes of allergic sensitisation in the UK"* [Humfrey 1996].

9.2 Conditions for mould growth and dust mites

There are many different types of mould species, the most common in UK dwellings include the genera Aspergillus, Penecillium and Cladosporium. There are other genera commonly found in dwellings and they all thrive under slightly different environmental conditions. However, all moulds require the following to grow.

- Mould spores: these normally enter the building with the air from outside which normally contains several hundred spores per cubic meter although the concentration increases in summer when external concentrations are considerably higher than internal concentrations, even if the property has mould growth.
- Nutrients: the slightest layer of grease found on all building surfaces is adequate for mould to grow.
- Temperature: different mould species thrive under different temperatures. Most moulds will survive in the range 0°C to 40°C, found in most buildings. Moulds will normally thrive better in warmer conditions. However, over the range of temperatures normally found in dwellings, 10°C to 20°C, there is little change in activity.

- Moisture: The availability of moisture is normally the critical factor determining whether mould will grow in a dwelling. Moulds extract moisture from the substrate they are growing on which in turn extracts moisture from the surrounding air. Therefore, both the surface properties and local relative humidity are important factors. Mould can grow on hygroscopic materials, such as wood and leather, at a relative humidity as low as 70% whereas on surfaces which do not absorb moisture, such as glass and ceramic tiles, it requires water to condense on the surface. This will occur at a relative humidity of 100%. For common wall coverings, such as painted plaster or wallpaper, the critical surface relative humidity for mould growth is 80%. If a relative humidity of 80% is maintained for a period of several weeks then mould will grow. However, conditions within dwellings are rarely stable, either because the external climate is varying, the heating system is being turned on and off or moisture is being introduced into the dwelling through the occupants breathing, cooking or washing. Few detailed studies have been undertaken into the impact of transient conditions on mould growth and, those that have, indicate that mould will grow if the relative humidity is above 80% for 50% of the time [Adan 1994]. Analysis of monitored data from dwelling livingrooms and bedrooms indicate that this corresponds to a mean relative humidity of 80% [Oreszczyn and Pretlove 1998].

Proliferation of the house dust mite (HDM) can be reduced by removing the source of food (human skin scale) by microporous barriers, killing mites with acaricide sprays, heat treatments, freezing or dehydrating mites. Controlling the relative humidity and temperature is the only long term cost effective method to control dust mite proliferation prior to sensitisation. The dominant HDM in the UK, *Dermatophagoides pteronyssinus*, rapidly increases its feeding, defecating, mating and egg production above 73% relative humidity at temperatures above 25°C. The HDM depends on mould to break down skin scales and make them digestible. However, if the relative humidity is above 85% the mould activity results in scales so decomposed as to make them inedible to the mites. Adult mites dehydrate and can survive no longer than 6-11 days at a relative humidity less than 50%. As temperature falls, egg output slows and egg-to-adult development time rapidly extends from 34 days at 23°C to 123 days at 16°C (at 75% relative humidity). This is one of the reasons why central heating has been suggested as a possible cause of asthma. However, the evidence for this is inconclusive and whereas the temperature associated with central heating may encourage mite populations to expand, central heating should substantially reduce the relative humidity therefore reducing mite numbers. Temperatures have to remain below freezing for some time in order to kill adult mites. Although many texts refer to maintaining an *absolute humidity* below 7g/kg, as proposed by Korsgaard [1983], this rule of thumb only applies to well-heated dwellings. For room temperatures below 20°C the relative humidity of surfaces with soft furnishings should be kept below 75% with occasional periods of a week or two below 50%. This is most easily achievable in dwellings during cold spells when the external air is very dry.

The above clearly identifies relative humidity as the key parameter for determining if mould or dust mites will survive. As relative humidity is a function of both the moisture and the temperature of the air the next sections identify the key factors that impact on these, both in general terms and specifically at the locations where mould and mites live.

9.3 Micro environmental conditions

Moulds and dust mites live on surfaces and within soft furnishings respectively and hence it is the microenvironments that surround these locations that are of importance and not just the average airspace conditions within a dwelling. This is particularly the case in older uninsulated properties so often occupied by the fuel poor.

A relative humidity of above 70% is often quoted [BSI 1989] as the critical condition for mould growth. This figure refers to the average conditions of the air that is typically found in dwellings where mould occurs. However, mould and house dust mites grow on surfaces and the local surface conditions can be significantly different from the air temperature. Since relative humidity is a function of both temperature and the moisture content of the air, the local relative humidity close to the surface of a colder wall will be higher than the relative humidity in the bulk of the room. This is particularly the case in dwellings that are only partially heated, uninsulated, or where thermal bridges occur, such as where insulation is missing or where there is two or three-dimensional heat loss [Oreszczyn 1992]. In order to keep the surface relative humidity below 80%, and therefore avoid mould growth, on an uninsulated brick wall the critical airspace relative humidity can be as low as 65%. For a well-insulated cavity wall the critical airspace relative humidity can be 75% and on a typical cold bridge it can be as low as 60%. For the above reasons there may be cases when the average room conditions are below the critical levels when either mould or mites should flourish but where there are specific microenvironments where moulds or mites can and do thrive.

9.4 Moisture production and movement in dwellings

The amount of moisture in the air within a room is dependent on the moisture produced within a space and the moisture transported into and out of the dwelling predominantly by ventilation.

MOISTURE PRODUCTION

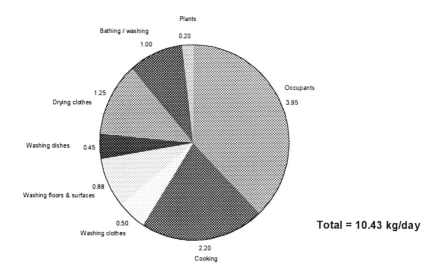

Fig. 9.1. Typical moisture production for a family of four (kg/day)

The amount of moisture produced and released into the air within a dwelling can range between 1 and more than 20 litres per day, depending on the number of occupants and their moisture production activities. Figure 9.1 shows a typical moisture production profile for a family of four. The largest moisture source is due to the perspiration and exhaled air of the occupants, followed by cooking and then bathing. However, personal habits can radically affect this. For example, microwave cooking of ready-prepared meals can substantially reduce the moisture associated with cooking. Also, occupants who spend all day inside the dwelling, such as retired couples, will result in increased occupant moisture production when compared to working couples. Dwellings should be designed to cope with a typical range of moisture production in a dwelling of a given size. However, any dwelling can experience mould growth if excessive moisture is generated. Dwellings should be designed to cope with some internal drying of clothes and if the dwelling is a flat without a secure garden or laundry facility it is likely that a substantial proportion of drying will be carried out indoors on radiators.

MOISTURE TRANSPORT

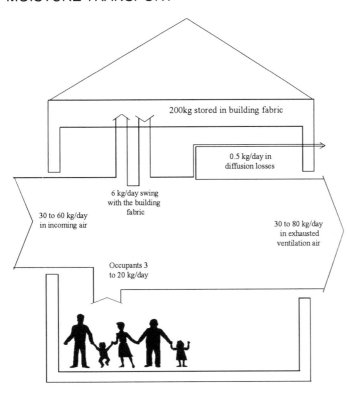

Fig. 9.2. Typical daily moisture flows in a dwelling

The predominant mechanism for moisture transfer in dwellings is ventilation. Figure 9.2 summarises the typical daily moisture flows in a dwelling. The incoming air has moisture associated with it and additional moisture is added to this air due to the moisture production detailed above. This moisture is then transported through the fabric of the dwelling by diffusion and by the extract air. This air has greater moisture content than the incoming air so increasing the ventilation rate reduces the internal moisture content. The moisture transfer through the fabric by diffusion is negligible in terms of the total quantity directly carried by the air. If this moisture condenses as it diffuses through the fabric then interstitial condensation may arise and the fabric may be damaged. Such interstitial condensation can occur if a wall is drylined with insulation but no vapour barrier is introduced.

Up to 200 litres of moisture may be stored in the fabric of a typical dwelling. As moisture is released into the space some will be absorbed into the fabric and then when the moisture production is stopped the moisture may then be re-emitted into the internal air. The extent of moisture absorption and desorption depends on the surface materials inside the building. An example of a highly absorbent environment would be a wood panelled room with very hygroscopic materials within such as a library. The other extreme could be a fully tiled bathroom, which contains minimal hygroscopic material. However, although moisture storage in the fabric plays an important role in short term fluctuations, what is absorbed is generally re-emitted and so this phenomena can be ignored if one is concerned with long term averages, as is the case with mould growth. The one main exception to this is where the fabric of a new dwelling is still drying out following construction. Evidence collected from a variety of new dwellings suggests that significantly elevated internal relative humidity is experienced for about 6 months to a year after construction is completed which can lead to mould growth.

9.5 Internal temperature

The internal relative humidity is not only dependent on the internal moisture but also on the internal temperature. The lower the temperature is for a given moisture content, the higher the relative humidity. Therefore, as dwellings become hard to heat and occupants cannot afford to heat them, the relative humidity rises within the dwellings. What is more, although ventilation may reduce the moisture content of the air, it will also reduce the temperature in a dwelling occupied by the fuel poor. In these situations, when the ventilation rate increases beyond a certain level the temperature drops and so the relative humidity rises. These two factors often compete against each other. Therefore, whereas a strategy of increased ventilation to eliminate mould growth may be appropriate in a fuel rich property, it may be less so for a fuel poor dwelling. This is indicated in Figure 9.3 below. In addition to the ventilation rate, the temperature will be affected by the type of heating system, its control, cost of operation and the levels of fabric insulation.

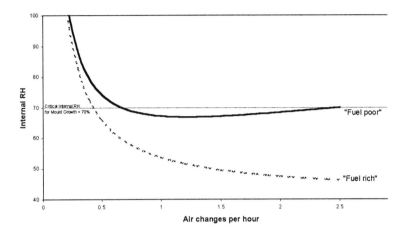

Fig. 9.3. The impact of increased ventilation on internal relative humidity

9.6 Relative humidity modelling

Because the internal surface relative humidity is dependent on both the moisture and the temperature, the relative humidity is dependent on many inter-related factors including the following:

- Fabric Insulation
- Occupants' ability to pay fuel bills
- Occupants' moisture generating activities
- Hours of occupation
- Number of occupants
- Heating system, efficiency, control and distribution
- Ventilation
- External climate

In determining the internal surface relative humidity, many of the above factors interact in a relatively complex manner. As a result of this complexity, a computer model called Condensation Targeter II [Oreszczyn and Pretlove 1998] has been created in order to gain an insight into the potential risk of mould growth in a particular dwelling with a standard occupant. This model utilises the thermal model BREDEM-8 to predict the internal temperature. Added to this monthly temperature calculation is a simple algorithm for calculating the surface temperature and a steady state moisture model [Loudon 1971]. Predictions of internal relative humidities for 36 dwellings, mostly modern, have been compared with measured relative humidities and

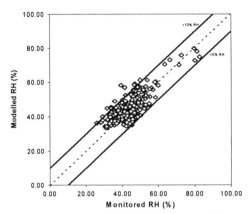

Fig. 9.4. Condensation Targeter II monthly modelled versus monitored relative humidity (Oct-May) for 36 dwellings

shown to agree to within +/- 10% for 95% of cases as indicated in Figure 9.4.

9.7 Sensitivity

In order to determine the impact that different factors have on the risk of mould growth, a sensitivity study using Condensation Targeter II has been undertaken. Figure 9.5 shows the impact of changing fabric, services and occupancy factors on the average air relative humidity within an uninsulated top storey flat.This flat is modelled as having an occupant who can afford to heat the property to 21°C for an average 10 hours a day and the standard case has a predicted relative humidity of 71%. The first bar shows the impact of changing the heating pattern. If heating is reduced to only 5 hours a day then the relative humidity increases to 92%. Alternatively if the flat is heated for a total of 16 hours a day then the relative humidity drops to 58%. The different interventions are ordered from left to right as roughly having less significance. Figure 9.6 shows the same dwelling, but this time the tenants are assumed to have a limited fuel budget of £60 per week. In this case insulation has as significant an impact as ventilation.

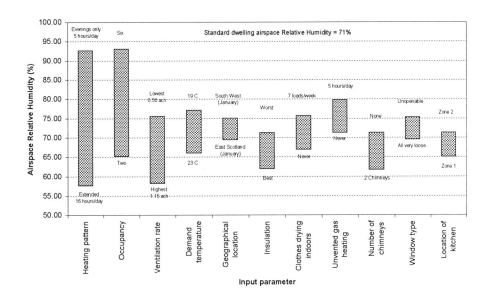

Fig. 9.5. Sensitivity study for uninsulated top-storey flat with two exposed walls and roof where fuel cost is no object

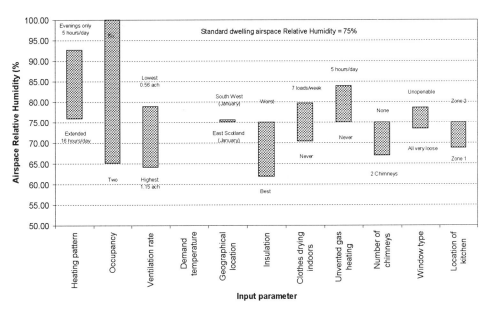

Fig. 9.6. Sensitivity study for uninsulated top-storey flat with two exposed walls and roof where fuel cost limited to £60/week

9.8 Mould Index

Chapter 8 proposes the introduction of an Affordable Warmth Index (AWI). A similar index has been developed called the Mould Index (MI). The Mould Index indicates the risk of mould occurring on the coldest surfaces in a dwelling. If the average surface relative humidity of the coldest surface equals 80% in any month then the MI will equal zero, whilst if the average surface relative humidity equals 50% in any month then the MI is given a value of 100. Between these two criteria the MI is assumed to be linear and can extend below and beyond the 0 and 100 MI ratings. A high value on the AWI scale indicates affordable warmth in a dwelling and a high value on the MI scale indicates *avoidance* of mould growth.

Using the computer model described earlier it is possible to examine the impact that different interventions have on the AWI and MI. It is also possible to model dwellings where fuel cost is limited and where it is unlimited, which has an impact on the MI values. As the level of insulation is improved in a poorly insulated dwelling, the AWI and the MI increases. These are indicated in Figure 9.7 below.

As expected, the fuel poor MI values, for situations where there is limited fuel expenditure, are lower than the fuel rich MI values where fuel cost is unlimited. Although the trend is similar, as the level of insulation in a dwelling reduces, mould growth is far more likely in dwellings of the fuel poor.

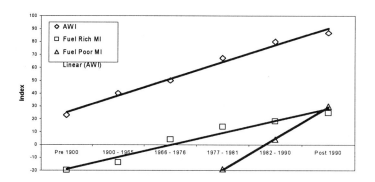

Fig. 9.7. Effect of insulation on AWI and MI

The impact of ventilation rate has also been modelled as shown in Figure 9.8. This indicates that if the ventilation rate is increased then the AWI gets worse. As the ventilation rate is increased the general trend of the MI is positive. However, in dwellings occupied by the fuel poor the

MI gets better until the air change rate reaches 1 air change per hour. Beyond this point the risk of mould growth increases significantly. This is shown in Figure 9.8 below.

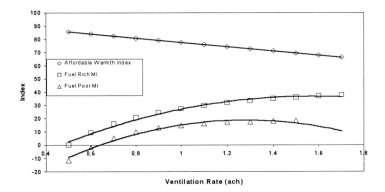

Fig. 9.8. Effect of ventilation rate on AWI and MI

9.8 Conclusions

There is increasing concern about the health implications of the indoor environment where we spend about 90% of our time. Since we spend the greatest proportion of this time in dwellings, it is essential that the environment within dwellings is healthy. At present many occupants live in conditions which are damp and considered unhealthy because they have high levels of mould growth or high concentrations of house dust mites. These properties are increasingly being refurbished and it is essential that the most appropriate strategies are undertaken to improve these properties. The impact of different refurbishment options not only depends on the existing fabric and services but on the occupant's ability to adequately heat their property, the moisture they produce and the way the occupants use the services. Measures, which reduce fuel poverty will normally also reduce the relative humidity and so may improve health. However, under some circumstances the mould index indicates that reducing the fuel poverty index may increase the risk of mould growth. The mould index presented in this chapter highlights one mechanism, Condensation Targeter II, by which the impact of all the main parameters relating to mould growth can be assessed. Although there is a range of far more complex tools which may have specific applications in interpreting the specific problems relating to a particular dwelling, for many properties the data required to operate these more sophisticated models with any improved accuracy is not available.

10

Winter morbidity and fuel poverty: mapping the connection

Janet Rudge

This chapter describes a proposed tool for evaluating the outcome of investment in affordable warmth. The development of this tool could also be of use to both local and health authorities in monitoring the effects of such investment. Ideally, it would be used to provide evidence of significant benefits in terms of reductions in costs to the health service, other services, or to building maintenance budgets. The case could then be made for further investment with assistance from a variety of different Government Departments or funding bodies.

10.1 Changing priorities

Britain has long had a reputation for cold, draughty and poorly insulated dwellings although the general public places a high priority on keeping warm. Even so, the relatively mild climate has often been used to justify the low priority given to energy efficiency in construction [Rudge 1995, 1996]. The Government now recognises the need for improving energy efficiency in buildings in order to limit their contribution to global carbon dioxide emissions and climate change. Legislation is in place relating to new construction, but the current rate of building replacement means that most people will be living in the existing housing stock, largely built before energy efficiency requirements were introduced, for many years to come [Boardman, 1993].

If measures are introduced to increase energy prices in order to limit energy use, this will adversely affect those who live on low incomes in buildings which are poorly insulated, with inefficient or expensive heating systems, and are therefore hard to heat. These are the 'fuel-poor' households, who are often the elderly. They are also the households least able to invest in energy efficiency so that savings could be made in the long term.

The Home Energy Conservation Act [HECA 1995] required that local authorities plan to reduce domestic energy use by 30% over a ten year period. The initial local authority bids had to include provision for the fuel poor but the Act does not make allowance for additional funding towards upgrading existing housing stock. HECA progress reports are submitted annually and they highlight targets with regard to the reduction in carbon dioxide emissions and energy use as well as the percentage improvement in energy efficiency. They do not relate specifically to progress with respect to fuel poverty so there is a danger that the specified targets will take priority. The main provision of the proposed Private Member's Bill on Fuel Poverty and Energy Conservation would have required local authorities to report on progress on dealing with fuel poverty. The New Labour Government has suggested that this data will be asked for on a voluntary basis, having negotiated withdrawal of the Bill in light of their proposed improvements to the Home Energy Efficiency Scheme [NEA, 1999; DETR, 1999]. The significance of action on fuel poverty will consequently remain unsupported by the force of legislation.

Justification for public investment normally requires demonstrable cost benefits. Potential energy savings resulting from increased energy efficiency can be calculated. The savings relating to carbon dioxide emissions are also measurable. In fact, the fuel poor have limited resources available for fuel expenditure, however energy-efficient their homes are made. In such cases, the benefit of energy-saving improvements is often 'taken back' as extra comfort [Milne & Boardman, 1997]. This means that there would be limited reduction of energy use or even none at all. This could militate against investment in hard-to-heat homes occupied by low-income households.

The indirect savings to the health service, or other social support services, are less easily demonstrated than energy savings. The cost-effective argument for investment will overlook the value of improvements to fuel-poor households because the benefits of providing affordable warmth, in terms of improved health and comfort, or reduced building maintenance costs, are difficult to quantify. If evidence were available to show that health targets were met as well as environmental targets, funding for energy efficiency measures may be justified from more than one Government Department. The emphasis would be shifted from that of purely saving energy towards the *rational use* of energy as recommended in the Watt Committee Report: *Domestic Energy and Affordable Warmth* [Markus, 1994].

The aim of the proposed mapping tool is to provide such evidence. A pilot study has been funded by the Eaga Charitable Trust to investigate the feasibility of this methodology, using the London Borough of Newham as its basis.

10.2 Housing and health

As described in Part One, Britain suffers a high rate of excess winter deaths and these are shown to increase as outdoor temperatures fall. This country's record is poor compared with many others with similar or even much colder climates. It has been argued that this is due to the relatively low indoor winter temperatures commonly found in British homes [Boardman, 1991].

Although the connections are widely recognised, demonstrating direct causal relationships between housing and health is problematic because of the many confounding variables which preclude an accurate comparative assessment of risk factors [Ambrose, 1996]. Dr Collins has described some of the complexities in Chapter 3 and they are further discussed in Chapter 18 by Dr Bardsley. One problem in researching the links is that self-reporting of illness attributed to cold and damp homes may be regarded as too strongly motivated by the desire for re-housing. There are also difficulties in funding longitudinal health surveys before and after interventions are made to improve housing conditions. Because they are expensive, these surveys usually involve relatively small sample populations. Some work has shown, however, that improvements in health have resulted from investment in energy efficiency in housing as described, for example, in Chapters 7 and 12.

10.3 Development of the methodology

For the purposes of developing an evaluation tool the intention was to use data which was already available to existing agencies and, by stepping back from the detail, to invent a relatively robust method of assessing health benefits of increased energy efficiency.

Since low temperatures are associated with certain illnesses - these have been identified as predominantly respiratory and cardiovascular disease [Kunst *et al.*, 1993] - it was thought possible that these types of illness could be shown to be more prevalent in winter among the fuel poor. In order to do this, therefore, it would be necessary to locate:

    a)   households on low incomes;

    b)   housing with poor energy rating, (where the combination of these two features would identify the fuel poor);

and c)   those who suffer from cold-related illnesses in winter.

These data would then be overlaid to look for correlations. Plotting at regular intervals could potentially reveal health improvement in the community following upgrading of buildings.

      Mapping provides a simple and accessible visualisation of the data analysis and it has other advantages as a tool in this context. GIS (Geographical Information Systems) for spatial data analysis is becoming more commonly employed in a variety of fields. It is already used by some local authorities and so, for them, the mapping tool could easily be accessed and tied into existing databases. GIS also functions as an epidemiological tool and the overlaying of health information with building and socio-economic data is therefore well served by geographical analysis.

10.4 An indicator for ill health related to cold

Previous chapters have referred to excess winter deaths as an indicator of adverse health effects of low winter temperatures. Curwen [1997] describes the definition of excess winter deaths as the number of deaths in the 4 month period from December to March, less the average of numbers in the preceding 4 months (autumn) and the following 4 months (summer). For the purposes of monitoring effects of home improvements, however, it is more appropriate to use a measure of excess winter morbidity, or illness. (Once death has occurred there is no longer any potential cost to the health service!) This may be calculated similarly, so the same 4 month winter periods are compared with the rest of the year as taken from August to July. This methodology proposes the use of a ratio (similar to the Excess Winter Death Index), which does not therefore rely on absolute numbers for its value.

      For the pilot study, it was decided that health data would be more easily obtained by using hospital admission statistics from a health authority as a single source. It would be more problematic attempting to gain information from a number of GPs across a borough, whose data collection methods probably vary and who may not all be prepared to make their records available. The East London and the City Health Authority provided records of hospital admissions over the four year period from August 1993 to July 1997 with 21 fields of data, including admission dates, type of admission, diagnosis and specialty codes, age and gender. (This range of data gives scope for analysis beyond the initial parameters of the pilot study, with a view to pursuing the work further.)

At first, it was envisaged that buildings and households could be identified at postcode level to show correlations between ill health and fuel poverty. However, due to the issue of confidentiality, the health authority was unwilling to give postcoded addresses as part of the data. Identification by enumeration districts, or 'e.d.'s, (these being census-related subdivisions of wards) was acceptable to the authority and it was thought would still give a reasonably detailed classification of data in relation to the area's population. (Newham constitutes 460 e.d.s for a population of approximately 220,000.)

On the other hand, it was recognised that when hospital admission statistics were broken down to enumeration district level (each constituting a population of approximately 500) the e.d. may not be a large enough unit to register significant variations. As this has proved to be the case, e.d.s will have to be clustered in the final analysis. This could mean analysis at ward level (there being 24 wards in Newham) or, possibly, grouping e.d.s according to building age and type, or energy rating. The method of zoning within the borough would be an area deserving further exploration.

10.5 Low income indicator

The question of which is the best indicator to use for low-income households is debatable. Some studies use a measure of deprivation based on points of information from census data. The aim of the proposed methodology is to use data that should be within the normal remit of collection by the local authority, as far as possible. The indicator should relate to ability to pay for heating rather than more wide-ranging deprivation factors (although some of these may be relevant to residents' state of health). The intention of this methodology is to be robust and is therefore concerned with statistics at a broad level. It avoids the level of detail required to look only at those individual households falling within the strict definition of fuel poverty as now used by the Government (a minimum of 10% of income needed to be spent on fuel to achieve acceptable levels of comfort).

Newham Council's published '*Poverty Profile*' [Griffiths, 1994] suggests that the receipt of Housing Benefit is a good indicator of where people are living in poverty. However, as Housing Benefit is not available to offset mortgage payments, the many elderly owner-occupiers of older terraced houses, with poor energy efficiency characteristics, would not be identified from this information. It is likely that they constitute significant numbers of the fuel poor in Newham, so this data would not be sufficient on its own.

Council Tax Benefit is available to all householders, including owner-occupiers, and does overlap Housing Benefit receipt. Details of those claiming this benefit would therefore be necessary together with, or possibly instead of, Housing Benefit data. These data were considerably more difficult to obtain, but may not be in future as councils rationalize their data collection systems. Alternatively, the indicator used could be that of Income Support claims, which the Local Government Anti-Poverty Unit can supply by ward as at 1996 and plans to produce by e.d. Benefit data is usually supplied as at the date of extraction so a historical record over a particular time period is unlikely to be obtainable by present data collection methods.

Of course it is not possible to include the many who do not take up benefits to which they are entitled, nor those whose incomes are still low but are just above the qualification cut-off point. So all data will underestimate numbers actually living on low incomes in relation to fuel expenditure necessary for comfort and wellbeing.

10.6 Housing energy performance

To assess the energy ratings for all properties in the borough, even on the basis of external surveys, would be a lengthy exercise beyond the scope of the pilot study. The Energy Report of the 1991 English House Condition Survey [DETR 1996] gives average energy ratings for typical dwellings which were derived from case studies. It classifies buildings according to the combination of age, ownership and type (low-rise or high-rise flat, terraced house and so on). In order to use these average ratings, house types in Newham were identified to match the case study examples as closely as possible.

The energy rating used is the Government's Standard Assessment Procedure, or SAP. This rating measures the cost of heating per unit floor area, based on heat loss characteristics of the building and the efficiency of the heating system. In some boroughs this information may already be available for dwellings at a basic level. Because of HECA, Newham Council had set up a database of residential properties by address and tenure, and energy rating details are gradually being entered from a variety of sources. The information already included within the borough was so far incomplete, particularly with regard to the privately rented and owned sector. However, the validity of the House Condition Survey average ratings in the Newham context could be tested using examples for which data did exist.

Information on the age of buildings was collated from the Planning Department to a single map, which allowed classification of each e.d. according to the predominant building age. From this map, it is very clear that in most cases whole streets have similar characteristics (one e.d. usually comprises two or three streets).

Computerised aerial photographic surveys of the area could help to identify recently constructed housing and to determine the storey heights and types of buildings. However, only a physical survey of the area was found to produce reliable confirmation of these details. Historical information relating to building improvements presents further difficulties. Newham Housing Department provided details of large scale Estate Action schemes carried out over the period under study. However, the energy rating improvement resulting from various upgrading works could only be estimated and records of smaller scale refurbishment works could not be obtained.

A spreadsheet was developed showing each ward so that combination details of building age, type and, where known, tenure could be entered as percentages for each e.d. These were then matched against the EHCS combinations to select average SAP ratings for the e.d. and ward, or other grouping of e.d.s.

10.7 Winter severity

Finally, from Meteorological Office data showing daily records of temperatures, wind speed, rainfall, and sunshine hours, the relative severity of each winter in the 4 year period could be determined, and the winter conditions compared with those during the rest of the year. Different measures of winter severity could be investigated for links with annual variations in hospital admissions for cold-related diseases. These could include the number of days with a mean temperature below 5°C, the number of days when the minimum temperature fell to 0°C or below, as well as the monthly mean and monthly mean minimum temperatures. As Curwen [1997] points out, variability of temperature, or wind chill, for example, could be part of the seasonal influences on health.

10.8 Conclusions

At the time of writing, the analysis of data for the anticipated potential correlations is incomplete. Preliminary work with figures for emergency hospital admissions indicates a slight excess winter incidence for all causes, a greater excess for ischaemic heart disease and a still greater excess for respiratory conditions. Various causes will also be analysed for different age groups. The method for grouping e.d.s within the borough and relating building information to health statistics needs further investigation.

It is clear that acquiring all the relevant data for this project was not without some difficulties but, on the other hand, officers of Newham Council and the Health Authority were extremely helpful and supportive of the work. On the basis of the work done, there would seem to be a good case for using existing information databases to produce a potentially useful mapping tool. When using databases held by others, there is no control over their quality, but once relevant queries are set up and these are used on a regular basis, they might be expected to become more reliable. There will always be problems over the issue of confidentiality when dealing with health matters and benefit receipts. If the data can be classified by enumeration district, it removes the possibility of personal identification. Address details do not include the e.d. in all databases, although it is possible to convert postcodes to e.d.s in GIS, with the loss of a small measure of accuracy.

Deprivation or low income indicators vary in usefulness, but work continues under different agencies to expand these and test their effectiveness. The variety of sources of information relating to buildings should become less of a problem as local authorities progress towards completing their energy databases.

As discussed above, it is likely that the pilot area selected will prove too small to find demonstrable correlations, but the work could be extended to include other boroughs in order to test it further, and in more detail, over a wider area. A greater variety of residential circumstances than is usual within one East London borough would probably be necessary to identify significant differences in health profiles and use of services.

If anticipated correlations between fuel poverty and high rates of winter respiratory illness are found, then the tool could be developed to estimate the attached costs to the health service to be set against the cost of building upgrades. Furthermore, it could demonstrate the value of increased energy efficiency as a preventative health measure as well as its potential contribution towards improved quality of life for the socially excluded fuel poor.

It would be a means of regularising the collation of data which is routinely produced separately by a number of departments within a local authority, or by different agencies, such as local and health authorities. In this way the summary of all the data would be made more useful to each organisation.

It should be possible to use the tool to monitor, over time, a number of different benefits resulting from energy rating upgrading. These could include, along with health improvements, reduced building maintenance costs and increased lifetime of the housing stock. The proposed methodology intentionally takes a broad view of conditions. It is not appropriate for assessing the detail of how far energy ratings are increased by particular measures, or their specific health advantages. With regard to the problems of damp in addition to cold in dwellings, the assumption is made that improving energy efficiency should include proper consideration of ventilation and relative humidity control.

However, in addition to the monitoring role, the mapping should highlight areas where problems combine to make fuel poverty likely, identify where certain health problems predominate and require servicing and indicate the types of housing that need upgrading most urgently.

Typical Newham housing mix

11

Modelling the health cost of cold damp housing

Jean Peters and Matt Stevenson

Introduction

The reduction of inequalities in health is a key factor in the current Government's health policy. At local level, Health Improvement Programmes are one of the mechanisms by which local priorities and targets are being identified and addressed. Further, the promotion of inter-agency working with partnership alliances and joint working between health authorities, local government and housing organisations is meant to be central to policy making. Housing conditions have been a feature on the public health agenda ever since the industrial changes of the nineteenth century with the resultant requirement for rapid provision of prolific but cheap housing for local factory workers. Poor quality housing has an effect on the health and welfare of its residents, although other social and economic factors are also important. Improvements in housing can potentially lead to better health and therefore a reduction of the burden on the NHS. However, in the current financial climate there is a need to indicate the cost effectiveness of any intervention. This chapter aims to quantify the additional level of use (over and above that for those living in warm, dry homes) made of health services, and the costs, by those living in cold damp housing in England, for health problems associated with the housing condition. This expenditure could potentially be saved were all houses to be made energy efficient.

11.1 Prevalence of cold or damp housing

In one of the largest surveys of housing conditions, 84% of 19,725,000 dwellings in England were found to have central heating [Department of

Environment, 1996]. However, in 8% of these houses the heating was inadequate in that it was not considered capable of maintaining a recommended minimum temperature of either 21°C or above in the living room or 18°C or above in the hall when outdoor temperatures were 1°C [Department of Environment, 1996]. It was reported that 1.2% of all houses had no form of fixed heating appliance. The issue is further compounded in that, even given the necessary heating facilities, it is estimated that there are 6.6 million households who cannot afford to heat their homes to a satisfactory standard [SCOPH, 1994]. Such households usually contain vulnerable groups. With outside temperatures averaging 2°C, the homes of 36% of unemployed and 31% of single pensioners were found to be heated up to 16°C only, and when outside temperatures dropped to freezing, 4.7 million people had room temperatures of less than 9°C [SCOPH, 1994]. With respect to damp, over 22% of householders reported problems with condensation and mould [Department of Environment, 1996]. However these figures may be an under-estimate. In Glasgow 28% of houses were affected, [Martin, Platt and Hunt, 1987], whilst a 25 - 50% prevalence was found in public sector housing stock in Edinburgh [Platt *et al.*, 1989].

Problems of cold and damp are not independent as, in general, damp houses are more difficult to heat and poorly heated houses are susceptible to damp. The twin problems of cold and damp result from an energy inefficient house which will have either high heat losses due to ventilation and lack of insulation or high energy input costs because of an inefficient or expensive to run heating system, or both. The energy efficiency of a building can be described in terms of its 'energy rating' (*see* Chapter 8 for further discussion). This is a measure of the annual unit cost of heating a building to a standard regime, assuming a certain heating pattern and specific room temperatures [Department of Environment, 1996]. One such energy rating is the Government's Standard Assessment Procedure (SAP) on a scale from 1 (highly inefficient) to 100 (highly efficient).

11.2 Health, use of health services and cold, damp housing

Associations between cold and/or damp housing and mortality [Curwen and Devis, 1988] and specific physical and mental morbidities are well documented, [for example, Hyndman, 1990, Hunt, 1993, Raw and Hamilton, 1995, Ambrose *et al.*, 1996]. However, the healthcare costs associated with cold or damp housing are less well documented. They include those associated with primary care consultations with a general practitioner or practice nurse and with secondary care for admissions, both planned or emergency. Pharmaceutical expenditure is a further cost. Costs will vary depending upon the induced condition and requirements for treatment. If the condition leads to death there will be no further health care costs for that individual. A systematic review from 1990-1998 of 22 databases, which covered health, social sciences, architecture, the grey literature and current research, found no information on use of health services by individuals for ill health associated with living in cold damp housing. Thus, whilst a person may suffer from ill health because of their housing state there is no quantifiable evidence that their use of health care and demands on health care services differ from that of people living in warm dry housing.

Nevertheless, some studies have attempted to estimate the health and health care costs of living in cold and damp housing. For example, the cost to the NHS from cold housing and excess deaths and morbidity associated with low temperatures in the home has been estimated at £220 million [Lawson, 1997]. Other studies, some of which looked at costs of health services for residents in poor compared with non-poor housing have come up with figures of £84.2m (1996 prices) [Barrow and Bachan, 1997] through £600m (1994 prices) [Carr-Hill, Coyle and Ivens, 1993] to £2.4b [National Housing Federation, 1997]. Given the lack of original research data and the considerable variability in published estimations, we have modelled the costs of selected key health conditions. These conditions have been identified from the literature as having a direct association with living in cold damp housing.

11.3 Modelling the costs

Those conditions used within the model, defined using the International Classification of Diseases, Injuries and Causes of Death (version 9) (ICD9), are shown in Table 11.1. In the ICD coding system each disease/ cause of death has an ICD number thus enabling consistency of definition internationally. Poor mental health is associated with both damp and cold housing. For the purposes of this modelling exercise, the costs have been calculated on the assumption that all additional mental health problems are attributable to damp housing only, as damp was found to be statistically significant, whereas cold was not [Hopton and Hunt, 1996]. This also avoids any potential double-counting as it is difficult to apportion the individual impact of cold and damp using existing published evidence.

There are a number of other accident-related health problems, and even deaths, that can be attributed to living in either damp or cold conditions although the association is less direct and these have not been included in this modelling exercise.

Table 11.1. Health conditions used in the model which are directly associated with damp or cold housing

Damp housing		Cold housing	
Disease	*ICD9 code*	*Disease*	*ICD9 code*
Asthma	493	Ischaemic heart disease	410-414
Allergic rhinitis	477	Heart failure	428
Bronchitis, bronchiolitis	466	Stroke	430-438
Mental health (non organic neuroses)	300		

A health service perspective was employed for the model. Only those costs associated with ill-health and cold damp housing and relating directly to health services are included. Costs incurred by individuals because of the health and housing condition are not included. General population prevalence rates for each health condition and specific prevalence rates for populations living in cold or damp housing were used (Table 11.2).

Table 11.2. Disease prevalence by damp housing condition and NHS spend per annum for England (1992/93 prices)

Disease	*Population prevalence*	*Prevalence in damp houses*	*Derived prevalence in dry houses*	*Estimated NHS spend*	*Estimated NHS spend (if all houses dry)*
	%	*%*	*%*	*£m*	*£m*
Asthma	4.9	5.8	4.1	424	408
Rhinitis	0.5	0.6	0.4	94	90
Bronchitis/ bronchiolitis	2.8	3.1	2.5	104	102
Neuroses	3.4	4.1	2.8	453	435

Sources: OPCS [1995], Department of the Environment [1996], NHS Executive [1996] Raw *et al.* [1996]

Given the lack of published information on health service use, particularly at the level of the individual, a top down approach was used. Costs (broken down by service: hospital, primary care, prescriptions, community care) were attributed to each disease, and apportioned according to disease prevalence by housing condition and prevalence of damp housing stock. The main cost figures were taken from national reference data sources [NHS Executive, 1996] and are expressed in 1992/93 prices. This source does not apportion all costs between disease areas, leaving some unallocated and it is thus expected that the figures calculated will be an underestimate of the true cost. Similarly, costs incurred within the private health services or associated with mortality have not been included (Tables 11.2 and 11.3).

11.4 General assumptions and limitations within the model

- The use of NHS resources per sufferer is equal regardless of housing status.
- There are equal numbers of residents per house, regardless of housing status in relation to cold and damp.
- The population are based in private residences and not institutional establishments.
- In disease areas that are not large and are grouped together with other diseases, e.g. bronchitis, the percentage of consultations atttributable to the disease will equate to the percentage of resources used.

11.5 Asthma, rhinitis, and bronchitis and bronchiolitis associated with damp housing

Data were taken from the English House Condition Survey [Department of Environment, 1996] and a health study of a subgroup from that survey [Raw et al., 1996]. From these two sources, the following were determined:
- the percentage of damp housing;
- the reduction in the number of damp houses, should they all be made energy efficient (proxied as a score of 60 on the Government's SAP scale);
- the prevalence of each disease in the overall population;
- the prevalence of each disease amongst those people living in the damp housing.

The reduction in the number of disease sufferers if all houses were to be made energy efficient, could thus be calculated from the disease prevalence in energy efficient houses. Assuming that there is no difference in resource use between those people living in damp housing and those who do not, this reduction can be turned into a cost estimate using simple proportions (Table 11.2). A SAP rating of 60 was chosen for this modelling exercise as this is the value associated with a medium-sized semi-detached house with a modern gas central heating system and above average insulation [Department of Environment, 1996].

11.6 Mental health problems associated with damp housing

The information used to estimate the health service costs of mental health problems due to damp housing was taken from a Scottish study [Hopton and Hunt, 1996]. It has been assumed that the increased percentage in the prevalence rate for those living in damp houses compared with those that are dry can be applied to England. The cost of neuroses has then been divided into those incurred by people in damp, and in non-damp housing. An estimation of the savings, if all houses were to be made energy efficient, can thus be made.

Table 11.3. Disease prevalence by cold housing condition and NHS spend per annum for England (1992/93 prices)

Disease	Estimated proportion of expenditure caused by cold energy inefficient houses	Estimated NHS spend	Estimated NHS spend (if all houses warm)
	%	£m	£m
Ischaemic heart disease	1.6	£800	787
Heart failure	1.6	121	119
Stroke	1.6	1,280	1,259

Sources: Collins [1995], Raw and Hamilton [1995], NHS Executive [1996], Sloan [1996]

11.7 Ischaemic heart disease, heart failure and strokes associated with cold housing

The following data sources have been used:
- the 20,000 excess deaths in the winter in the UK [Collins, 1995]

- the proportion of the population of England to that of the UK [ONS, 1994]
- 30% of excess winter mortality due to cold housing was through ischaemic heart disease [Sloan, 1996]
- 50% of excess winter mortality is due to cold housing [Raw and Hamilton, 1995]

The following assumptions have been made for modelling purposes:
- the relative mortalities of cardiac failure and stroke to ischaemic heart disease remain constant during the winter period
- the ratio of morbidity cases to mortality cases remains constant throughout the winter period
- 5% of the entire cold housing stock is energy efficient. This figure was used instead of zero to acknowledge that a minority of houses are expected to be cold due to reasons other than energy inefficiency.

11.8 Estimated cost savings

Table 11.4. Annual cost savings by disease if all houses were made energy efficient

Disease	Savings £m (1992/93 prices)		
	Central estimate	Lower estimate	Upper estimate
Asthma	16.72	9.45	23.99
Rhinitis	4.23	2.39	6.07
Bronchitis/bronchiolitis	2.51	1.42	3.59
Neuroses	18.20	10.29	26.11
Ischaemic heart disease	12.93	7.35	19.06
Cardiac failure	1.96	1.12	2.89
Stroke	20.69	11.76	30.49
Total	**77.24**	**43.78**	**112.20**

Table 11.4 gives the estimated cost savings that would be achieved were all houses to be made energy efficient, plus lower and upper limits. The values used to calculate the upper and lower limits are detailed in Table 11.5. These estimates assume that the prevalence of a health condition in previously damp or cold houses would immediately return to the prevalence found in warm and dry houses. This assumption is likely to be

slightly inaccurate, as there is likely to be a time lag before the two prevalence rates equate. In an extreme case this may take until the next generation, where all children are raised in energy efficient houses.

The sensitivity analyses show that the estimated savings are likely to be in the region of £43.78m-£112.20m per annum (1992/3 prices). The central estimate is £77.24m.

Table 11.5. Assumptions for sensitivity analyses

	Upper estimate
Lower estimate	
• the percentage of damp houses caused by poor energy efficiency was reduced from 23% to 13%	• the percentage of damp houses caused by poor energy efficiency was increased from 23% to 33%
• the percentage of cold houses that were energy efficient was increased from 5% to 10%	• the percentage of cold houses that were energy efficient was decreased from 5% to 0%
• excess morbidity caused by cold housing decreased from 50% to 30%	• excess morbidity caused by cold housing increased from 50% to 70%

11.9 Limitations of this modelling approach

In many respects the estimates in this modelling exercise are conservative.
• They do not include the prevalence of those with undiagnosed disease but consuming health resources.
• They only include costs and resource use for one component in those diseases such as mental health where both, damp and cold, have a role. Further, they do not allow for any additive or multiplicative effects of both cold and damp on specific health conditions.
• The list of diseases costed within the model is not exhaustive. Costs of other diseases and general ill-health, both directly and indirectly associated with cold and damp, have not been included.
• The reference source used for the costings does not apportion all costs between disease areas, leaving some unallocated. It is expected that some of these unallocated costs are attributable to the diseases modelled. Likewise, pharmaceutical expenditure or costs of community care for some of the diseases modelled are not available. Whilst such costs might be small they are unlikely to be negligible.
• Indirect costs, such as costs of sick leave, are not included.

- No costs for mortality have been included.

There may also be some over-estimating, in that:
- The prevalences of diseases in damp conditions were based on information obtained following a prompt [Raw *et al.*, 1996]. This might have resulted in bias.
- The mental health problems may not be caused by the poor housing conditions but instead those afflicted may have to reside in low quality housing.

11.10 Comparisons with other published estimated costs

Other published estimates of costs of health care associated with cold damp housing are the defined but fairly crude calculations of Lawson [1997] and those of the National Housing Association [1997] which have no supporting explanation. Lawson's rough calculation of NHS costs for morbidity and mortality attributable to a 5°C drop in winter temperature was £220 million per year. This estimate only covers costs of morbidity and mortality associated with cold and not that associated with damp so in one respect it may be an underestimate. However, the figure used of 8,000 deaths per drop in temperature of 1°C applies to a drop in temperature below the average air temperature for the time of year. This is then multiplied for a drop of 5°C and produces a figure of 40,000 deaths. Other researchers' estimates for death each year in Britain, as a consequence of the winter conditions, range from an average of 20,000 per year [Collins, 1995], through 30,000 aged over 65 [Sloan, 1996] to 40,000 for the winter of 1995/96 [Aylin, 1999]. The other sources of data include two primary research studies but both have small sample sizes of 203 adults, and 107 households, in Carr-Hill, Coyle and Ivens [1993] and Barrow and Bachan [1997] respectively. An extrapolated upper cost figure from Carr-Hill, Coyle and Ivens findings is £2bn per year as an NHS spend trying to treat conditions caused by poor housing. Only a portion of this £2bn can be attributed to the effects on health of cold damp housing although the size of that portion, whilst unknown, is substantive.

This modelling exercise has focused on a limited number of selected health problems but which have a clearly documented association with cold and/or damp housing. As such it is likely to be an underestimate of the costs involved. However in the absence of other published data it provides an indication of the possible size of the problem and an indication of the potential improvements in specific health conditions and savings in the cost of health care provision for these were all these houses made more energy efficient. At £44.23m-£113.36m, this equates to approximately 0.2% of the total UK NHS expenditure for 1992/93 of £36,261m. It must also be borne in mind however that remedial action or provision of warm and dry housing may not necessarily solve all health problems associated with cold damp housing as other social and economic factors may be implicated.

11.11 Recommended methodologies to guide policy

A number of housing interventions have recently been carried out or are underway at the present time. However the opportunity to quantify health service use is not being taken in most cases in what are robust intervention studies. Within the overall costs of such studies the acquisition of self-reported data on use of health care services would be virtually negligible and even the additional resources needed to monitor general practice and hospital records comprise a small component of the total cost. The opportunity to develop a sound database of robust evidence should not be missed. Given the current focus on health improvement programmes and reductions in inequalities, sound evidence is essential for guiding the development of appropriate targeted interventions. However, whilst small local interventions can collect very useful data and make an impact on the health of local populations, such information would be of much greater value if all interventions followed similar protocols and agreed to collect a core set of outcome measures. If the information obtained is to be used to inform policy it needs to be generalisable to the wider population within the United Kingdom. Finally, in the planning of any future interventions to improve health, whether focussing upon housing or much broader issues, it is essential to ensure that all specific factors that impact upon health, such as the geography, economics, demography, housing variations in stock and type, are considered at the development stage.

There is little published information on the cost to the NHS associated with living in cold damp housing. This modelling exercise provides a useful tool for estimating the additional costs borne by the NHS due to cold or damp housing, and the potential savings that could be made were all houses energy efficient. In this study, the estimated savings are at least £43m with a central estimate of £77m. Hard evidence is needed to validate the modelled data and for informing future policy decisions.

The authors would like to acknowledge the support of the Building Research Establishment for the work described in this Chapter.

Part Three:

Inter-agency partnership in practice

12
Asthma: lessons of the Cornwall Housing
Intervention Study

13
Treating cold, damp and asthma with affordable
warmth

14
Promoting partnership: the Nottingham example

15
Urban care: working in partnership with
communities

16
Working in partnership: lessons from ten case
studies

12

Asthma: lessons of the Cornwall Housing Intervention Study

Ian Mackenzie and Margaret Somerville

12.1 Introduction

It has been widely accepted [Acheson 1998] that there is an association between cold, damp housing and respiratory complaints. There has, however, been surprisingly little research on the effect of interventions to improve housing on the health of the occupants. In part, this lack of research may reflect the difficulties of applying biomedical research designs, such as randomised controlled trials, in this area, and of co-ordinating the inter-professional and inter-agency working necessary to ensure that an appropriate study design with an effective intervention is both adequately funded and successfully completed. It is also vital to keep the people at whom the intervention is directed involved in the study. In this chapter we discuss a research project carried out in Cornwall [Somerville *et al.*, 1999] which illustrates many of the problems faced by researchers in this area. We hope that many of the lessons learnt during the work in Cornwall will help future researchers in this area to plan studies and evaluations more effectively. The project has been recognised by a National Health Service "Beacon Award" for health improvement.

None of this work would have been possible without the enthusiasm and co-operation of the Housing Departments of the local authorities and housing associations within Cornwall. The present Government has emphasised, through national strategy, the importance of partnerships between health services and local authorities in addressing the broad determinants of health and working together to produce a local Health Improvement Programme [DoH 1998,1999]. This project would not have happened without that joint working.

12.2 The Cornwall Housing Intervention Study

In autumn 1994 the abolition of the South and West Regional Health Authority led to the distribution of reserve funds to Health Authorities throughout the region. Cornwall and the Isles of Scilly Health Authority received £1.2 million of non-recurring funds to spend before March 1995. The Authority had endorsed the Director of Public Health's annual report [CISHA 1994], which returned to the issue of housing and health first highlighted by Dr. Sharp, Medical Officer for the City of Truro in 1896 [Truro 1896] and awarded to the district councils in Cornwall £300,000 to use on housing improvements to improve health outcomes. Each of the six district councils in Cornwall received £50,000. The funds were transferred from the Health Authority to the District Councils using section 28A of the 1977 NHS Act relating to the transfer of public finances between public authorities. Only houses within public ownership were able to receive grants. One district council subsequently transferred the ownership of the public sector houses to a housing association.

To use NHS money in this way was a leap of faith in the absence of published evidence of effectiveness of such an intervention. The transfer of funds was not without criticism. A member of the Local Medical Committee commented *"it is more important that every penny possible is spent on the acute Trusts where there is a severe shortage of funds and beds"*.

CHOOSING THE TARGET GROUP

In evaluating this innovative use of NHS money to improve health outcomes, as required by the Health Authority's auditors, a number of issues arose. Many aspects of poor health have been associated with poor housing conditions, including respiratory problems, arthritis and rheumatism, mental illness, infectious diseases and accidents [Lowry 1991, Bardsley *et al.,* 1998]. These conditions reflect different aspects of poor housing, such as cold, damp, poor design, over-crowding and poor lighting, and different groups of the population, such as children and the elderly. The choice appeared to be either to improve a group of houses currently in poor repair, and evaluate all aspects of health in those houses' occupants, or to target a group with known health problems and improve houses individually. An attempt to measure all possible health outcomes relevant to housing conditions was considered impractical. It was thought more useful to identify a single client group for whom there was reasonable evidence of a link between their illness and their housing conditions.

In choosing which client group to target, it was essential to consider the prevalence both of the health problem in question and the associated aspect of poor housing. In common with other areas of the country [Bardsley et al., 1998, SHCS 1996], Cornwall has a high proportion of damp housing (around 25%). Asthma has been a local health priority in Cornwall, as it is the commonest chronic disease of childhood.

The association between damp housing and asthma is plausible, as damp housing favours mould growth and provides ideal living conditions for house dust mites; both mould and mites contain allergens which can trigger asthma attacks. As this association is discussed elsewhere in this book, it is not dealt with further in this chapter. Evidence for a link between housing conditions and asthma comes from surveys which have consistently reported an association between damp housing and respiratory symptoms in adults [Brunekreef 1992] and children [Platt et al., 1989, Dales et al., 1991, Cuijpers et al., 1995]. Some have relied solely on self-reporting of symptoms and damp and mould [Brunekreef 1992, Dales et al., 1991] while another has independently verified reported housing conditions [Platt et al., 1989]. When lung function has been objectively measured, however, the association has not been confirmed [Cuijpers et al., 1995, Strachan and Sanders 1989] and children from damp homes do not have worse bronchospasm than other children, despite increased parental reporting of symptoms [Strachan 1988] Case-control studies have confirmed an association between respiratory symptoms and damp and mouldy housing conditions in primary school children [Verkhoeff et al., 1995, Williamson et al., 1997], but not in children under 5 years old [Lindfors et al., 1995] or of unspecified age [Leen et al., 1994] or in adolescents [Strachan and Carey 1995].

Despite these findings and the plausibility of the association, no intervention studies on housing had been published on whether it is possible to reduce the morbidity due to asthma by reducing the amount of dampness and mould in the house. The public health medicine department in Cornwall therefore proposed to the district councils that the money should be directed at reducing damp in the homes of children with asthma.

CHOOSING THE STUDY DESIGN

A framework for evaluation had not been developed prior to the decision of the Health Authority. Although a briefing paper had been prepared for the Health Authority on the association between housing and health it was not expected that the Health Authority would decide to transfer funds.

Once the decision was taken there was then a pressing need to develop an evaluation framework which could determine whether the investment in housing led to an improvement in the symptoms and health of the children living in houses improved by the grants. We quickly realised that it would have been preferable to have developed an evaluation proposal to have been considered by the Health Authority at the same time as the grants were made available to the local authorities. This would have allowed a properly funded evaluation to be carried out. As it was, the evaluation was carried out by members of the Public Health Department and the housing officers of District Councils and Housing Association without any additional resources.

The importance of planning evaluation early and setting aside sufficient resources has been recognised by the Department of Health; those Health Authorities who have become Health Action Zones have been required to set aside between 5% and 10% of total initiative funding to carry out planned and purposeful evaluation.

Given the funding constraints for evaluation, the possibilities for the study design were limited. Medical interventions, particularly drugs, are normally assessed using a randomised controlled trial, but this design is not easily transferable to other settings such as housing. A proposal that children with asthma living in damp poorly heated housing should be randomly allocated to an intervention or control group was not thought feasible. The local authorities wished to improve houses with the worst conditions where children with the most troublesome asthma lived and there was a general feeling that to deprive half of these children of a warm dry home was not politically acceptable. There were also concerns that co-operation from the people allocated to the control group would be poor, as it would be impossible to maintain any form of subject blinding to the nature of the intervention. A two-phase study, with some houses being improved before others, was proposed, but there was pressure to use the funds quickly and so the suggestion was rejected. The study then had to proceed with a "before and after" design.

CHOOSING THE OUTCOME MEASURES

In deciding how to measure change in health status, both generally and in asthma-specific terms, we were again constrained by the time in which the study needed to be undertaken and the resources needed to measure the outcomes. There are several aspects to the assessment of asthma severity; the definition is dependent on the demonstration of variable airflow obstruction [BGAM 1997] and it is important to consider both the objective assessment of airflow obstruction and the subjective effects of the disease on reported symptoms and lifestyle. We were particularly interested to know if there were changes in use of health services and medication and the amount of time lost from school due to asthma. In addition, we hoped to assess change in general health status using the SF36 questionnaire.

As well as assessing health status, we needed to know that the improvements carried out by the local housing departments resulted in an improved indoor environment. We were most interested in the reduction of damp and mould and improved indoor temperatures, but would also have liked to assess levels of house dust mite, tobacco smoke, pet dander and nitrogen oxides [Bardsley *et al.*, 1998, Platt-Mills 1994], all of which have been implicated in triggering asthmatic symptoms.

Our only resource in assessing outcomes were the housing officers of the local councils, who needed to visit each property selected for improvement in order to decide exactly what work needed to be carried out. They agreed to complete questionnaires with the occupants of each house on the Health Authority's behalf. As the housing officers did not have any clinical training, it was not possible to ask them to carry out objective measurements of airflow obstruction, such as peak flow rates, and it was unreasonable to expect them to record details of medication. We therefore confined our health status assessments to a symptom-based outcome measure [Steen *et al.*, 1994], which asked about the children's respiratory symptoms in the month before interview, questions about the use of health services and time lost from school. The housing officers also visually assessed the state of the houses for damp and mould and took appropriate measurements to calculate a summary statistic on energy efficiency, the NHER [NES 1996], which was our main outcome measure for housing improvement. The NHER scale ranges from 0 (very poor) to 10 (excellent); the standard for new-build housing is 8. (Energy ratings are explained more fully in Chapter 8.)

In view of our inability to obtain information on medication through the questionnaire undertaken by the housing officers, we sought permission to consult the occupants' GPs for further information at a later date. When funds to carry out this part of the study became available, through a grant from EAGA, practice nurses were asked to extract the information on drug dose, frequency and contacts with health services from the children's medical records. As asthma symptoms and severity can vary markedly by season, we wished to collect information on health service and medication use for a year before the intervention took place and a year afterwards to allow for seasonal variation.

IDENTIFYING THE CHILDREN

Although many GP practices have a register of asthmatic people within their practice, such registers are not readily available or comparable on a county-wide scale and do not necessarily record information such as housing conditions. We did not think it was feasible to use such registers for identifying children. We considered, but rejected, suggestions of advertising for suitable candidates, on the grounds that we did not have the resources to assess a large number of applicants. A number of children with asthma living in damp council housing were known to the housing departments, but, because the current system of assessing people for re-housing on medical grounds did not rate such children highly, it was thought unlikely that the departments knew more than a small proportion of suitable candidates. Health visitors, asthma liaison nurses and paediatricians were therefore asked to recruit likely children. As the study progressed, it received some local publicity and news also spread by word of mouth. Families started to contact both the housing departments and the Health Authority directly to ask if they could be considered. A few were recruited in this way, but by the time knowledge of the study was widespread, the funds had been fully allocated.

Confirmation of moderate to severe asthma in the child was obtained from the patient's general practitioner, paediatrician or an assessment by the paediatric occupational therapist. A few children with respiratory disorders other than asthma (for example cystic fibrosis) were included where these children were awaiting housing improvements and were seen as a priority group.

CHOOSING THE INTERVENTION

The exact nature of the intervention was the subject of much discussion. The only direction from the Health Authority was that the intervention should produce as warm, dry and energy-efficient a house as possible, given the available resource. Housing departments needed to balance providing some improvement for a reasonable number of houses with individual houses which needed other substantial improvements. There was broad agreement that the main intervention would be the installation of central heating, but departments varied in the amount of insulation and ventilation which had already been installed in council-owned properties. Gas central heating would have been the installation of choice in most cases, but many parts of Cornwall have no access to mains gas, so alternatives needed individual discussion between the housing departments and the families involved.

It was anticipated that the funds available to each district council were sufficient to carry out the necessary improvements in about 20 households; that is, about £2,500 could be spent on each of 120 houses in the county.

12.3 Effects on housing, symptoms and school attendance

The results of this study have been submitted for publication[2] and are briefly summarised here. Of 138 households initially identified as possibly suitable for intervention, baseline surveys took place in 104 and 98 subsequently had an intervention, at an average cost of £3061 per house. Most of the houses were semi-detached or terraced and before the intervention had been heated by a coal fire or direct heating appliance. Gas central heating was installed in nearly half of the houses, with over a third receiving electric storage heaters and the rest having either solid fuel or oil-fired central heating.

Energy efficiency, as measured by the NHER scale, improved significantly in the 62% of houses for which 2 readings were available (from 4.4 to 6.5 on the NHER scale, $p<0.001$). Over 90% of children's bedrooms were unheated before intervention and over 60% were damp. Following improvements, only 14% of children's bedrooms were unheated and 21% were damp. All respiratory symptoms were significantly reduced after intervention; the greatest reduction was seen in nocturnal cough. Before intervention, most children had coughed on most nights in the month before interview, but afterwards, most children had only coughed on one or a few nights in the previous month. After the intervention, school-age children lost significantly less time from school for asthma in the 3 months before interview (9.3 days per 100 school days before intervention and 2.1 days afterwards, $p<0.01$), but not for other reasons.

12.4 Effects on health service and medication use

At this stage after completion of the intervention, we were able to identify 75 of the original 98 households who received an intervention, but less than half of these gave consent for us to examine their medical records.

Overall, there was a reduction in the number of contacts with NHS staff in primary and secondary care after the intervention, but there was a small increase in the cost of prescribed medicines. An economic analysis of these data and the housing costs found that the savings to the NHS exceeded the annual equivalent cost of the housing improvement.

This saving arose mainly as a result of the reduction in admissions to hospital and this reduction was heavily influenced by the reduced use of hospital by one individual.

12.5 Interpretation of findings

There is obviously a need to be cautious in interpreting these results. Without a control group, it is not possible to assess the contribution of several important confounding variables. Firstly, there is likely to be a response bias because it was not possible to blind the subjects or the assessors. In the first part of the study, the results depend on the parental reporting of symptoms and loss of time from school. The families were obviously aware of the intervention, and given their interest in receiving improved housing conditions it is probable that there is considerable bias in the reporting of symptoms. In the second stage, those households with children experiencing the greatest gains were probably the most likely to respond. As the response rate was low, the reported changes cannot be generalised to the whole study population.

Secondly, there is an inevitable selection bias, due to the need to identify children with asthma living in damp council housing. As there was no definitive register of such children, we are unaware of how representative this group is of all children with asthma or of all asthmatic children living in damp housing.

Thirdly, asthma in children changes as the children grow older and symptoms also vary substantially by season. Asthma symptoms and attacks are more frequent in the autumn and winter months and again for a short period in the hayfever season. The increase in symptoms in the winter months is linked with increased community infection with viruses affecting the respiratory system. Seasonal fluctuations in asthma severity may have influenced the results, as repeat questionnaires were not completed at the same time of year as the original. The time periods for collection of data were consecutive, however, and the size of the change in contacts with the NHS considerable, suggesting that the natural history of the disease cannot entirely explain the improvements seen. We would not expect to see substantial remission in asthma over the relatively short period of our study.

Fourthly, the treatment of asthma by primary healthcare teams has steadily improved in recent years. The rise in prescribing costs is accounted for by the increased use of preventative treatments for asthma. This increased use of preventative medicines is likely to have contributed to improved symptom control and a reduction in NHS contacts. A randomised controlled trial would be necessary to take account of these changes in management.

Fifthly, alterations in exposure to allergens from furred or feathered pets and passive smoking may also have contributed to the improvement. There were high levels of both parental smoking and pet ownership in households, with changes in both following intervention. While the reduction in parental smoking may have reduced children's symptoms, there was an increase in pet ownership which may have increased them. As we were not able to measure the levels of these triggers for asthmatic attacks directly, it is only possible to speculate on their actual influence.

Finally, many of the families were lost to follow-up. This particularly affected the health service use section of the study; although verbal consent had been given previously, less than half of the families responded to our letter seeking written consent to examine their medical records. In retrospect, we should have obtained written consent at the time of interview; the poor subsequent response may reflect loss of interest on the families' part. Council house tenants, particularly people with young children, are a mobile population and 12% of families moved within one year of the housing intervention, before the follow-up questionnaires were carried out. Co-operation with follow-up was difficult with some, although others were delighted enough with the results to speak to the media.   The issue of mobility raises the question of whether the intervention was worthwhile, if the child and family were only able to benefit for a few months. This needs to be taken into account when considering the benefits of the housing investment.

12.6 Economic considerations for the occupants

Further substantial financial benefits accrue to the residents themselves. This is represented as a potential financial saving resulting from a reduction in fuel bills. Although it is possible that households might have chosen to maintain household temperatures at the levels pertaining before the improvement, economic theory would predict otherwise. We would expect that a reduction in the real cost of heating would result in increased household temperatures. The evidence on the reduction in the use of the NHS services supports the view that people chose to improve the climate within their homes, rather than simply use the savings from more efficient heating systems to switch expenditure to other areas of consumption.

The results of this study suggest that the costs of installing central heating to improve the health outcomes of children with asthma living in damp poorly heated houses are less than the benefits to the NHS in terms of reduced contacts, particularly in-patient admissions.

12.7 Policy considerations

To use NHS money in this way was controversial in the absence of published evidence of effectiveness of such an intervention. Many public health interventions, however, have been undertaken without a strong evidence base and other considerations besides the quality and strength of the evidence, such as humanitarian issues, inform such decisions. Is evidence of effectiveness of social interventions on health outcomes necessary before suggesting that children with asthma live in decent houses? While such evidence may not be thought necessary before improving housing conditions, we should still seek to extend the evidence base and quantify the benefits for occupants.

The recent report on health inequalities [Acheson 1998] recommends the introduction of policies to improve the quality of housing, which the government has now recognised in its revised health strategy [DoH 1999]. The present government has carried on the steady transfer, commenced by the previous administration, of publicly owned houses into the private sector through housing and residents' associations. There is considerable concern amongst the Society of Mortgage Lenders about the housing fabric in private sector ownership. Anecdotally, the district councils considered that there were worse problems in the private rented than in the public sector, but we were unable to make our funds available to this group. People on low incomes who have bought their own homes have problems with maintenance and heating costs. Poor housing is increasingly becoming a personal rather than a public problem.

New opportunities are arising for the inter-agency working needed to make the required changes in current practice. Local Authorities now have a duty of partnership to work with Health Authorities and others to develop local health improvement programmes [DoH 1998, HSC 1998], which should address wider issues of health as well as health care. Health Action Zones also provide scope for addressing these issues. While establishing joint local priorities may aid local action on poor housing conditions, there also needs to be more co-ordinated central action linking the private and private rented sectors into these programmes. Organisational boundaries should not act as barriers to implementation. Nor should the NHS necessarily be seen as the source of funds for health-related housing improvements. We need a national strategy which acknowledges the links between poor housing and poor health and makes appropriate provision to tackle it.

12.8 The future

There are considerable methodological problems with this study, resulting from both its context and the intrinsic problems of carrying out this type of research. Houses and people are complex. Fitting central heating does not mean that it is used. Houses are static and people are mobile. Since its inception, we have regarded it as a pilot study and hoped that our efforts at evaluation would help others elsewhere to develop more methodologically robust studies. Nevertheless, as the study has progressed, we have felt that the findings have merit, despite the need to confirm them in other settings.

More work is now in progress. Also in South West England, Torbay Council have agreed to use their housing improvement funds in one area where the local tenants have surveyed their houses and found high levels of respiratory symptoms and damp housing. The houses will be improved over a period of 3 years, giving the local Health Authority and Plymouth University the opportunity to evaluate both the changes in the indoor environment of the houses and the health status of the occupants. With substantial NHS research funding, it has also been possible to measure more objective health and environmental outcomes. The stepped-wedge design of this study allows for comparisons between improved and unimproved houses each year, which should provide more substantial evidence of the link between poor housing and poor health.

Acknowledgements

Particular thanks and recognition are due to our research colleagues David Miles, Ken Buckingham, Pat Owen and James Bolt. We are also extremely grateful to EAGA for funding part of the evaluation.

13

Treating cold, damp and asthma with affordable warmth

Rob Howard and Roger Critchley

13.1 Background

At the time of writing, this study is not yet complete but, nevertheless, provides an example of how partnership working can bring huge additionality to a project, both in terms of resources (revenue and capital) and also expertise and knowledge. The project began with an application by NEA (National Energy Action) for Nottingham Health Authority funding which is administered through the Nottingham Health Action Group. NHAG is a multi-sector group (including local authorities, voluntary sector, and health professionals) that works strategically to promote the health of the people of Nottingham health district. It aims to tackle inequalities in health and identify and address the environmental causes of ill health. Nottingham Health District covers Broxtowe, Nottingham City, Rushcliffe, Gedling and Hucknall. NHAG has brought together a Local Agenda 21 health strategy - 'Health in your Environment', local health priorities and Health of the Nation targets. In order to fund projects which will contribute towards its aims, a budget was established - the Health Initiatives Budget.

NEA's Nottingham Office was established in 1995 as a demonstration project to examine innovative ways of tackling fuel poverty. The original application came about as a result of contacts made with Nottingham Health Authority through NEA's membership of 'Health in Your Environment' (*see* Chapter 14). This original bid, entitled 'Energy Housing and Health' aimed to reduce cold related illness for people experiencing fuel poverty. The aim was to target the most vulnerable households, and prevent these illnesses by providing good quality energy and condensation advice, and a package of heating, insulation and ventilation measures. From the lessons learnt from this, we hoped to be in a position to influence the maintenance and improvement policies of social housing providers when dealing with tenants at risk of cold and damp related illness. The application was successful and the project implementation began.

13.2 The Study

At this time, NEA's Senior Projects Officer, Rob Howard, met Roger Critchley (First Report) an independent housing and energy consultant and part of the Health and Housing Group, at an Eaga/NEA seminar. Tadj Oreszczyn from the Bartlett School was then brought in for advice and all partners agreed to work on the project together. As a result of this collaboration the project was hugely improved. In particular there had been little technical or research background to the project and only limited health monitoring was included. It was agreed that seven additional dimensions be built into the project:

- a critical look at different humidity control strategies;
- a high degree of technical input in relation to planning the improvements;
- before and after monitoring of temperature and humidities;
- before and after examination of house dust mite levels as well as the health of the occupants;
- a range of further monitoring techniques examining housing conditions and health of the occupants;
- detailed examination of costs;
- writing up the project results as a very detailed case study with full discussion, together with the provision of guidance drawn from the results.

This partnership, as well as improving the quality of the study, brought additional funding to further increase its scope. The five main funders then became:

- Nottingham City Council - for heating systems, heating controls and other physical improvements to the homes;
- Nottingham Health Authority - for insulation and ventilation measures;
- EAGA charitable trust - for technical support and project design and preparing and writing up the final report;
- Bartlett School of Graduate Studies/EDAS - for staff time;
- NEA - for staff time.

The overall aim of the project was narrowed down, and more specific objectives were defined. These included providing a detailed analysis of the most practical methods of reducing house dust mites, lowering relative humidities and improving the condition for fuel poor asthmatic tenants in Nottingham. In particular, it was hoped that we could reduce relative humidities to an average of 50 - 55%. The project aimed to act as a regional demonstration of the health benefits of improved housing. The main objectives were to:

- reduce the relative humidities in a small number of homes currently rented from Nottingham City Council by low income households where household members suffer from asthma;
- monitor the effectiveness of improvements carried out, both in terms of relative humidity, house mite populations and improving health of the occupants;
- detail the necessary planning and technical considerations in order to make effective improvements and achieve low running costs whilst balancing the need to keep capital costs to a reasonable level;
- explore in detail the possibilities, costs and difficulties in terms of achieving low relative humidities in low rise domestic housing;
- present the results of the project to social housing providers and other relevant bodies.

The initial operational partner was Nottingham City Council who identified households due to receive full central heating on the grounds of ill health (asthma). Additional criteria were also used to reduce the significance of other factors such as households with cats and/or smokers. After a number of initial contacts seven households were selected and the implications of taking part were fully explained over a number of occasions. In fact, the tenants themselves were regarded as key participants in the study, and fully involved and consulted throughout each stage. The potential benefits of participating were clearly explained as well as the expected level of disruption and a relationship was established which meant that we had enormous co-operation from them. We then undertook a detailed monitoring exercise over a one week period during February 1998 which included:

- NHER energy auditing
- fuel readings
- condensation survey
- air pressure testing
- humidity and temperature data logging
- house dust mite sampling

- SF36 assessment (self completion health questionnaire)
- asthma drug use
- lifestyle diary

Following the analysis of these results, the intervention was then designed for each household. This was a package of measures which included:
- well controlled full gas central heating system;
- insulation measures determined by the energy audit;
- a ventilation strategy which varied from single extract fans in bathroom and kitchen, to 3/4 house heat recovery ventilation system.

One year after the initial monitoring period, the exercise was repeated.

13.3 Partners' roles and responsibilities

The success of the study depended upon clear communication between all partners so that every one clearly understood their roles and what was expected of them. Communication took the form of several meetings between operational partners, progress reports to NHAG, and regular and lengthy face to face contact with the tenants. The work programme was divided as follows:

- NEA - project co-ordination, project design, contact with tenants, NHER auditing, SF36 assessments, liaison with NHAG;
- Bartlett School - air pressure testing, project design, house dust mite sampling, results analysis;
- Health & Housing - project design, condensation analysis, lifestyle diary, report writing.

There was considerable overlap and exchange between these roles, which was made possible by effective day to day communication.

13.4 Results

Although the results are currently being analysed, anecdotal evidence already suggests significant health gains by a number of the tenants who report verbally a reduction in symptoms and drug use. However, the health gains were only part of the study, and a number of lessons have already been learnt to be included in the final report. These include:

- the importance of strong and clear specifications when commissioning improvements and ensuring effective qualitative monitoring and enforcement of these specifications. In a number of cases thermostats and programmers had been set incorrectly by the installer, and radiators and controls had been placed in inappropriate places. For example, we discovered:
  - programmers set twelve hours incorrectly, so the heating came on at night, and went off during the day;
  - hot water cylinder thermostats set to 80°C instead of 60°C, wasting energy and increasing the risk of scalding;
  - radiators placed on internal walls rather than under the windows;
- the need to include energy advice as part of the installation contract. None of the tenants had received advice on how to use the new systems or been left with any literature to help them. When we visited some time after the installation, almost all were using the systems inefficiently. One of the tenants had altered the settings on the 3/4 house heat recovery ventilation system and another was worried about the cost of running a humidistat controlled extract fan. With appropriate advice, none of this need happen.

Following detailed analysis and reporting on the results, it is planned to publish a set of three good practice leaflets aimed at housing, health and energy efficiency professionals. These will draw on the key recommendations which come out of this study and link to other relevant studies in this field. The expected outcomes of the project are:

- improvements to the health of the tenants and a reduction in fuel bills and improved management of energy use within the homes involved in the study;
- indirect benefits through environmental improvements (e.g. reduced CO_2 emissions) resulting from the physical improvements;
- focusing of local attention on the inherent dangers of certain energy efficiency measures (e.g. excessive sealing of housing leading to inadequate ventilation) in relation to the health of the occupants;
- provision of information on the practical considerations which need to be taken into account when any landlord, building professional or other party undertakes improvements;
- highlighting those areas where there is a lack of technical, biological or medical knowledge and to inform the process of providing energy efficient homes for the fuel poor in Nottingham, without the dangers of house dust mite infestation.

13.5 A partnership - real or imagined?

When any two or more institutions or organisations, be they statutory or voluntary or individuals, join together on a particular project, it can be claimed that a partnership has been formed. Indeed this definition is used by the DETR in New Deal for Communities where they describe a partnership as "the organisation or group of organisations which will be engaged in the development and implementation of the Delivery Plan." [DETR 1999]

Closer analysis of existing "partnerships" reveals that, just as in personal relationships, the power relationships between "partners" are often unequal, the goals disparate and the experience can prove very unsatisfactory to one or more parties. The term partnership may, in many instances, be a euphemistic title used for publicity, funding or self-seeking purposes. This section examines in more detail the quality of the partnership in this project and the benefits that the partnership has, or may have brought to the different parties.

It is argued here that for a partnership to be effective five criteria should be met:

- each organisation is satisfied with its role;
- each organisation is able to work positively with others to achieve the defined project goals;
- each organisation brings different or additional resources to the project;
- there is a satisfactory system of communication between all organisations;
- there is an absence of power conflicts, role confusions, undermining of other partners or territorial disputes.

13.6 The relationship in the partnerships

The basic relationships of six main parties in this project are illustrated in the diagram overleaf.

The lines of communication in this diagram show there are unequal relationships between the various parties. Power and control has been unevenly distributed. The funders considered the original proposals and funding and then have not interfered with the day to day management of the project. The operational managers have all had responsibilities in different ways to their funders and have had to ensure they met the criteria of the funding application. However, the main role for the operational managers has been to discuss and decide on the exact details of remedial measures and the scope and implementation of all tests. At the same time a recording and evaluation function has been carried out

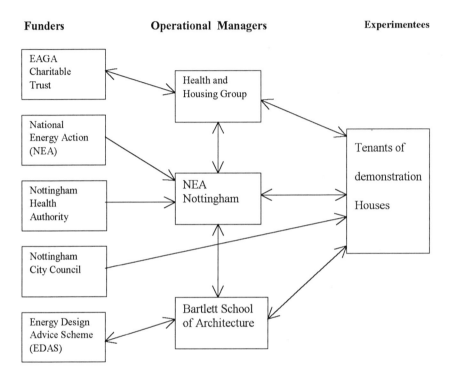

The tenants have been consulted throughout but a range of options has not been offered to them nor have they been involved in the planning process, the test results or the conclusions. In that sense their level of involvement can be considered to be low. In many partnerships this could be considered an important negative factor.

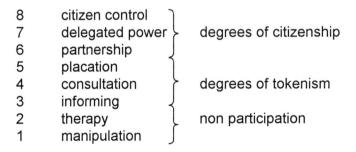

Fig. 13.1 Arnstein's ladder of participation

Arnstein suggests there are eight levels of participation and the ladder of citizen participation can be described with eight rungs [Arnstein 1969]. These rungs are helpful in placing the tenants' involvement in a context (Fig. 13.1).

Judged against this ladder this project involved consulting and informing the tenants about the process. Consequently, they cannot be judged to be equal partners in the process. However, whilst accepting that with considerably more resources a closer involvement could have been constructed, the actual basis of the project must not be forgotten, namely to improve the indoor environment and the tenants' health through the most appropriate technical measures. In a demonstration project such as this, neither an equality of partners nor an empowerment of the residents may be desirable or achievable. The considerable time spent explaining the process and background to the tenants and valuing their involvement has been a key feature, we believe, in ensuring high levels of co-operation.

13.7 Funding partners

At a funding level there could be considered to be a funding partnership. However, each funding organisation made their decision to back the project in isolation. There were no joint meetings of the funders or discussion between the different parties. In that sense the funders have been largely passive partners whilst the project has been undertaken. However, prior to its inception each funding organisation would have been actively involved in deciding whether or not to support the project.

Moreover, two of the funders (EAGA Charitable Trust and NHAG) have been able to call on the operational managers to feed into seminars and discussions on the issues of fuel poverty, health and housing. The funders will also be able to draw their own lessons and experience from the project when it is completed. However, there is one further desirable outcome. If the funders, or others who learn of this project, value the positive outcomes from the project, then future inter-agency discussions or partnerships may take place. The project has focused on a very important area where health authorities can work in partnership with social landlords and can be put into context with other ongoing and recent work e.g. the Cornwall initiative, Liverpool Health and Housing study, and the Southampton and BRE studies on house dust mites.

It is also important to recognise that each funder has brought to the project resources that assist in different ways:

- Nottingham City Council providing capital works;
- Nottingham Health Authority providing resources for ventilation and insulation works;
- EDAS providing technical expertise and advice and monitoring;
- EAGA Charitable Trust providing time to assist with technical aspects and recording;
- NEA providing the original ideas and staff time for implementation.

Against this background the funding partnership has been exceptionally positive with different agencies bringing different resources and expertise. Moreover the initial funding agencies (Nottingham City Council and Nottingham Health Authority) have provided the opportunity to encourage other funding participants. The jigsaw of funding has been productive, each agency bringing their unique contribution.

13.8 Operational partnership

At the centre of the partnership was the relationship between NEA, the Bartlett School of Architecture and the Health and Housing Group, each represented by one individual. The implementation of the whole project has relied on this small ad-hoc executive group. Due to its size, meetings have been minimal, contacts largely made by telephone and little time wasted on unproductive functions such as larger meetings of every interested party.

Some of the skills and experience of the three executive officers overlapped, for example, energy efficiency, energy audits and a keen interest in health and housing. However, other areas of expertise were confined to particular partners, e.g. research techniques, blower door tests, house dust mite analysis, choice of ventilation strategies. This has undoubtedly enriched the project to date. The project was open to a large number of approaches. For instance, the lead expert could have been a medical practitioner, a building surveyor, an energy assessor, a ventilation engineer, or an environmental health consultant. Whilst the final approach did not involve a medical practitioner, the range of disciplines in the operational management has been wide and has noticeably increased the value of the project.

In this case and where other health and housing issues are at stake, the value of a multi-disciplinary approach cannot be underestimated. Of course, such an approach can bring striking differences of opinion and arguments as to direction, but the benefits are considered likely to outweigh the disadvantages. At the heart of this project is a triangular relationship between asthma, fuel poverty and housing. In the Social Exclusion Unit's report, 'Bringing Britain together: a national strategy for neighbourhood renewal', a criticism of past government practice is made: *"a joined up problem has never been addressed in a joined up way"* [Social Exclusion Unit 1998]. It is hoped that the effective supportive partnership of funders and active partnership of operational managers in this project has addressed the health problems in " a joined up way".

As the structure of primary care changes new opportunities for partnerships open up and it is likely that some of the most effective partnerships will be forged at Primary Care Group level. We think there are many opportunities to work at this level as well as with health authorities to explore practical environmental solutions to health problems by looking closely at temperature and relative humidity.

14

Promoting partnership: the Nottingham example

Helen Thompson

"Working in partnership" is a term used throughout countless local and national documents. There are people meeting to work in partnership up and down the country with varying degrees of success. The Nottingham Health Action Group is a group of professionals in the Nottingham Health District who have made a commitment to doing just that! (Voluntary sector members are included here, as professionals who may or may not be paid for their contribution.) This chapter provides an account of multi-sector work in Nottingham which will improve the health and environment of people in Nottingham Health District through energy efficient and warm homes.

14.1 Health in your environment' strategy

The Nottingham Health Action Group (NHAG) was established in 1996 and works across the Nottingham Health District which covers five Local Authority Districts. NHAG aims to:-

'work strategically to promote the health of the people of Nottingham Health District, tackle health inequalities and identify and address the environmental causes of ill health'.

In February 1997, NHAG launched a Local Agenda 21 Health Strategy for Nottingham Health District called 'the Health in Your Environment Strategy', with the blessing of the five Local Authorities, the Health Authority and the Voluntary Sector Forum which inspired the work. Another major part of the work of Nottingham Health Action Group is to contribute at a local level to the Government's 'Our Healthier

Nation' initiative which focuses on reducing the major causes of ill health and inequalities in health.

Nottingham's Public Health Directorate took the lead in convening the Group, which involves senior people from the five local authorities, the Health Authority, the voluntary sector, primary care and health trusts and is part of the formal Joint Commissioning structure within Nottingham Health District.

The Voluntary Sector Forum picked out four themes from the Health in your Environment Strategy:- food, energy efficiency, asthma and transport. Their proposal for NHAG to organise four major seminars about these topics was agreed and members of NHAG identified key players from their sectors to join the planning groups. They were all co-ordinated by the Health Action Group Officer, on behalf of NHAG and were called the 'FEAT' seminars.

14.2 Planning for success

The effectiveness of each seminar was due to the way they were planned. Because the planning groups were multi-sectoral from the start, the priorities, concerns and good practices of each sector could be taken into account at an early stage. They comprised people with expertise in the subject and those who could influence people within their own sector. They were one day events which included key speakers with national and local perspectives, discussion groups focusing on how improvements could be made by each participant, displays and information sharing opportunities, and plenaries at which action areas are shared.

The common aim running throughout all of the seminars was to bring people together with health, social and environmental perspectives to identify current problems, causes and potential solutions, to share information about good practice and to motivate action to overcome the problems. Linking local with national initiatives was an important part of each event. The planning groups met on a number of occasions and membership changed as plans progressed to suit the topic.

14.3 Outcomes from the Energy Efficiency, Health and Environment Seminar

The aim of the Energy Efficiency, Health and Environment seminar was:-'to improve the health and environment of people in Nottingham Health District through energy efficient and warm homes'. This seminar took place during Warm Homes Week and was attended by 77 people from a varied backgrounds. They heard from a wide range of authoritative speakers, including Health Authority representatives and the local Member of Parliament, who is Chair of the Parliamentary Warm Homes Group. The Director of Newark and Sherwood Energy Agency described the Council's substantial improvements to housing and the health benefits this has brought and Peter Ambrose outlined a research project on the health effects of bad housing conditions (*see* Chapter 17 in this book).

Delegates divided into six discussion groups which each began with a short presentation, then developed into discussion which generated a number of action points aimed at improving the local situation. Three of these points were selected for feeding back to the whole seminar at the end of the day. The groups discussed :-

- best practice;
- energy advice awareness – signposting, basic information and grants for improving homes;
- benefits to people's health of energy efficiency improvements and what health services can do to improve energy efficiency;
- improving private sector housing, energy saving and healthier living initiatives;
- energy efficiency and vulnerable groups – general;
- energy efficiency and vulnerable groups – older people.

There were numerous outcomes from the discussion groups but key points were raised by several groups as follows:-

- the need for awareness-raising training to emphasise links between health and energy efficiency among health workers, e.g. through Primary Care Groups;
- an inter-agency framework should be developed to ensure co-ordinated action and training – including the voluntary/community sector, local authorities, primary care groups, emergency services;
- cold and damp related illness should trigger action for energy efficiency improvements to patients' homes – tackling the problem at source through inter-agency co-operation;
- the health sector (NHS) should play a role in funding energy efficiency initiatives, eg capital improvements to housing, and in supporting external programmes promoting such measures;
- Nottingham Health Action Group, Nottingham Health Authority and partners should lobby for appropriate legislation, e.g. the Warm Homes Bill;

- waiting rooms in, for example, GPs' surgeries and hospitals, should be used to promote energy efficiency and provide information;
- existing resources should be directed to those in greatest need.

14.4 Enabling joint work in Nottingham Health District

A combination of circumstances, policy framework and the political will at a national and local level made it possible for this joint work to happen. The policy framework was provided by a push at international level, in the form of Local Agenda 21, to balance environmental, social, and economic factors to make a better life for all. In addition, the introduction of the 'Our Healthier Nation' agenda was directed at reducing the major causes of ill health and inequalities in health. A national agenda through current guidance in all sectors, together with a local history of commitment to partnership work between health, local authorities and voluntary sectors, laid the foundations of the multi-sector Nottingham Health Action Group (NHAG). With the development of the Health in your Environment Voluntary Sector Forum (HIYE), came excellent ideas for development which NHAG could take forward and Public Health were willing to resource a temporary post to co-ordinate developments.

14.5 Potential barriers to progress

Progress in all multi-sector work can be blocked through a lack of understanding or willingness for people to work together. This can include the attitudes of professionals, such as cynicism, negativity, arrogance, or suspicion of people from "outside" the organisation. 'Professionalism', traditional career pathways and different agendas, both personal and organisational, can make it easier for staff to 'go with the flow' rather than risk unpopularity. People from different organisations may have difficulty in understanding each others' perspectives and could find it easier to work with others who share a common training, outlook, or way of working. The importance of joint work is not always understood and much lip service is paid to working in partnership. Joint work is creative: it does not always fit with existing protocol and new protocols may need to be established to allow new initiatives to happen.

There is sometimes a tendency to start writing a new strategy once the last has been completed without implementing the first. Employees can lose sight of what they are paid to do; for example it is important for staff to play a part in developing the organisation that they work for, but this can take up time which detracts from the objective of the work. Staff overloaded with statutory work may feel that they have no time to carry out preventative work or to review their working practices. A 'heads down' or 'get our own house in order' mentality can prevail, which may block real progress in multi-sector work.

14.6 Overcoming difficulties

In this case, multi-sector planning groups were established from the start. These identified key issues across the board and allowed for the development of seminar aims common to all partners, to which all could contribute. Many planning meetings were held and, although this meant longer preparation time, rewards were gained through the sharing of expertise and contacts. The vision was combined with looking at how ideas could be implemented since both vision and practical implementation are needed for ideas to work.

Trust was developed between individuals in different organisations through determined effort, sharing of credit and retaining a sense of humour. Both local and national experts contributed. For example, a representative from the Centre for Alternative Technology in Wales participated, partly due to links with a local voluntary organisation. Finally, dedication and determination to make things succeed on the part of the co-ordinator and the planning groups were an essential part of this process.

Ideas for overcoming barriers to effective joint working are discussed further in Chapter 16.

14.7 Building on success

NEA have produced a report about the seminar, compiled by Tim Gray for Changeworks and the Health Action Group Officer, which is available nationally as an example of good practice.

Nottingham's Joint Consultative Committee received an initial report about all four seminars and endorsed the work that was carried out. The Chairman ran a follow up meeting to all four seminars to review progress. It will be the responsibility of the Nottingham Health Action Group and the seminar participants to ensure that the ideas developed through these seminars are turned into action. Some proposals have already been incorporated into Nottingham's Health Improvement Programme document, which means that there is a commitment locally to implementing those ideas.

Nottingham Health Action Group will be taking the lead on developing the 'Affordable Warmth' workstream as part of the local Health Action Zone. A key objective is to train Health and Social Services staff, who already visit people in their own homes, to identify basic home energy efficiency problems. Once they have the householder's consent, they can refer them to a home energy efficiency expert who will put the wheels in motion to make improvements.

One project developed was a 'Beacon of Excellence' capital bid funded through regional 'Our Healthier Nation' monies. Age Concern Nottinghamshire is working on a pilot project together with a local Primary Care Group to make improvements to older people's houses in a locality within Nottingham city. Primary Care Groups, which are developing currently, are interested in taking many ideas forward. The signs are promising that the Nottingham partnership experience will lead to improvements in the health and the home environments of the local community.

15

Urban care: working in partnership with communities

Adrian Jones

15.1 Urban decline

Birmingham, like many other parts of the country, is witnessing a deterioration in the condition of its housing stock which numbers 390,000 dwellings. Although the effects of this can be seen amongst all tenures, nowhere is it more evident than in the private sector, where home ownership has increased by almost 20%, between 1981 and 1991, to a total of 225,000 dwellings [OPCS, 1991]. The way a home is maintained is a critical factor in the health of the occupants and the long-term viability of properties and neighbourhoods. However, householders face many barriers to taking the necessary action to protect their home environment. Surveys indicate that there is a widespread lack of information, advice, skills and resources that are accessible to householders of all tenures.

Currently there is very little slum clearance and associated redevelopment within Birmingham and, therefore, the city's ageing housing stock is likely to stand for a considerable time. At the same time, reducing public funding for house renovation grants can only provide resources for dealing with the worst properties and this is not sufficient to keep up with the scale of housing disrepair.

At the end of the millennium Birmingham is likely to have in excess of 100,000 privately owned homes that are in need of urgent repair and many more that are rapidly deteriorating because of lack of routine maintenance and improvement [Bucknall Austin, 1994]. Local house condition surveys have reflected national findings [DoE, 1996] which demonstrate a high level of housing disrepair and indicate low home energy ratings for properties. Within the Sparkbrook constituency in Birmingham a 10% sample of 15,000 households revealed that 63% of homes suffered with disrepair, whilst only 11% had undertaken any remedial works during the past 12 months. Of those dwellings in disrepair the major problems related to cold (49%) and damp (47%) [Birmingham City Council, 1998].

Both locally and nationally, underinvestment in the maintenance of homes leads to deteriorating housing conditions and consequently a risk to the residents' health and safety. There is also a financial implication as neglected smaller items of disrepair develop into larger and much more expensive problems. In addition, increasing amounts of money can be wasted in trying to heat properties that suffer from over ventilation, an inefficient heating system or inadequate insulation. The ultimate consequence of a lack of home maintenance is a general decline within a whole neighbourhood leading to a deteriorating environment, poor health, reduced community confidence and increased crime (*see* Chapter 17). This cycle of decline is well recognised as a precursor to slum clearance or other dramatic local authority intervention to regenerate an area. These solutions are extremely expensive, but the cost is not the only issue as urban decline brings trauma and ill-health to families and social breakdown to communities.

15.2 An alternative approach

Birmingham's urban care strategy aims to avoid this scenario. Traditional approaches have not been working because housing policy generally only addresses the extremes of slum clearance or area renewal. Both of these options have very expensive unit costs and affect relatively few properties. The challenge is to establish new services that can prevent urban decline. This is being done by involving residents in developing new approaches to the maintenance of their homes. Even in a climate of reducing public funding for house renovation, urban care is seen as a way of expanding activities that deliver on the Housing Department's objective of "improving housing in the city". Urban care fills a gap in housing policy and is potentially an effective and efficient way of protecting a large proportion of the housing stock and improving people's living conditions. The "stitch in time" principle is accepted as a cost-effective way of maintaining homes and neighbourhoods.

In developing the urban care approach, Birmingham is clear that the City Council, acting alone, does not have sufficient resources and cannot have a wide enough impact on house conditions. From the beginning the local authority has seen its role as that of an enabler. Although providing some maintenance services in-house, in many circumstances it empowers people and encourages the development of self-help initiatives within communities.

Householders' views have been sought on their housing problems and why it is that necessary works are not being undertaken. It has become clear that a narrow definition of home maintenance is not useful. What a local authority may categorise as home maintenance, home safety, energy efficiency, security and local environmental issues are not considered separately by householders. For example, a badly maintained gas fire may be a home maintenance matter but if it begins to discharge poisonous carbon monoxide fumes into the air it also becomes a health and safety issue. It is certainly an energy efficiency and fuel poverty issue as non-renewable energy and low-income householders' money is squandered. Similarly, failing to regularly paint a wooden framed window could cause a frame to rot and warp, which allows excessive heat loss from a building. It has also become apparent that "messages" on home maintenance topics get confused in residents' minds, such as sealing a house to stop heat loss whilst maintaining adequate ventilation to control condensation dampness. Any strategy needs to consider a house as a whole and not in a piecemeal way as can occur through the work of disparate specialist projects. Such an inclusive strategy has the benefit of providing a framework for agencies and organisations to work together more effectively with residents. Certainly it is important that there is joint working with the health authority as it is apparent that most home maintenance issues have a health connection. Perhaps one of the areas where this is most apparent is in relation to cold and damp living conditions.

An important aspect of the urban care approach is the way in which communication takes place and services are delivered. The public's view is quite hostile to many organisational initiatives that appear big and glossy but leave the lives of ordinary people untouched. "What difference did that project make?" is a question frequently asked by the people who are supposed to be most affected by an initiative. Organisations with energy efficiency, home maintenance and health messages can easily find their messages lost in a climate of information saturation and junk mail. Experience suggests that talking to people individually is effective but no organisation can resource this on the scale required and most fall back on traditional methods of communication such as leaflet distribution. Community communication networks may address this problem by effectively carrying valuable information for those last few metres, past a resident's front door. In addition, services take on a new relevance when workers are employed to work in neighbourhoods for projects managed by representatives of the community.

15.3 Consultation

Birmingham is a multicultural city and contains areas of poverty, poor health, high unemployment and older people living alone. With any consultation on the barriers people face in carrying out home maintenance, it is essential to involve all residents, including those who are socially excluded. The urban care approach depends on good consultation since services are designed to be delivered through residents' own community self help schemes. If the consultation exercise does not identify the real problems, it is unlikely that any resulting action plan will receive any degree of community support. The consultation also identifies viable solutions to overcoming barriers to action. Over a dozen methods of consultation have been employed, depending on their relevance to an area. Community conferences and workshops, including events for specific socially excluded groups, have worked well as a catalyst to local action. Alternatively, consultation displays and interviews have been conducted from a caravan parked in a resident's own street. Members of the urban care development team have even sat on garden chairs on street corners in order to enable a dialogue with people who just "don't have the time" to make their views known through more structured consultation. Apathy is the norm and so must be planned for.

As well as involving residents in the consultation exercise they have also been empowered to carry out their own consultation. One example of this is an assisted completion of a questionnaire in residents' homes with ethnic language support. Community groups came together to plan and design an urban care needs survey. Fifty local people were trained to complete the questionnaires and a survey of 1,500 households was undertaken with the support of a specialist survey consultant. The results were shared with the six community projects involved and also with a much wider group of partner organisations. The information was provided on CD ROM and written reports tailored to specific neighbourhoods. Here people had some ownership in identifying the barriers and so were more likely to be committed to their own local action plans and any subsequent community self-help projects.

The barriers to home maintenance that are most commonly identified in consultation exercises are:
- lack of knowledge
- lack of skills

- fear of "cowboy" builders
- lack of community resources
- negative feelings about area (linked to crime and poor environment)
- poor communication
- lack of money
- expectation of grant aid.

Many residents expressed the view that alongside the physical measures to tackle urban decline the strategy should also take into account the lack of community spirit.

15.4 Urban care strategy

The consultation resulted in a strategy that includes three elements: care for the home; care for the environment; and care for people. Its five objectives are to:
- prevent the deterioration of houses
- improve health, safety and security within the community
- reduce the consumption of non renewable energy in domestic premises and tackle fuel poverty
- prevent a decline in the local environment
- inform, empower and fully involve people affected by urban care.

This is achieved by:
- producing and distributing information resources to raise awareness
- providing an advice service
- improving access to appropriate training in a community
- enabling the employment of builders
- developing and supporting community self-help schemes
- enabling residents to finance necessary works in their home
- establishing a community communication system
- encouraging and enabling young people to play an active role in urban care activities
- identifying and celebrating success in a community.

15.5 Partnership

The information from a neighbourhood consultation is brought together to produce a local action plan. This is where partners from the public, private and voluntary sectors may act to support the community. Existing partners include the health authority, council departments, housing associations, police, fire service, colleges, schools, Age Concern, Victim Support, retailers and builders. Partners provide a range of support for any proposed project, which may include advice, funds, materials, accommodation, transport, accountancy or training. Developing partnership schemes can be complex and is often compared to a jigsaw where everyone has to be in the right position before the picture of the whole project can be seen. Patience and diplomacy can be tested at various stages along the way, particularly where larger organisations come to terms with real community empowerment.

The next step may be opportunistic, recognising the need for moving to the action stage quickly to maintain local interest and gain community confidence. Alongside this, longer term planning is required to ensure that an area's needs are addressed and that partnership activities are sustainable. Community schemes are developed which provide local people with the necessary information, advice, skills and resources to maintain and improve homes and neighbourhoods. Examples of community schemes are provided below, many of which are managed by locally elected residents' groups. Many of these initiatives have formed part of a government good practice guide [DETR, 1997].

It is recognised that there is no one 'community' in Birmingham and that neighbourhoods are made up of many communities. However, expressions such as 'community self help project' are used here to describe people coming together in a spirit of equality to manage and deliver services available throughout their neighbourhood. The expression 'home maintenance' is used to include energy efficiency, home safety, security, environmental improvement as well as routine structural maintenance.

15.6 Information, advice and practical assistance

The initial task of raising public awareness and providing information on how to maintain a safe, warm and secure home environment is most important. On their own, information leaflets have a limited value but, if designed properly, may be effective in linking initiatives. Successful publications have included information on how to carry out DIY works, which has been linked to the urban care training courses, and information on employing a builder which has been linked to accessing lists of reliable contractors. Publications have been written from the householder's perspective to ensure that they serve a useful purpose. Sometimes this is achieved by providing the funding to a residents' group so that they can commission their own information resources. In any case leaflets are market tested amongst their target audience before general release.

Information is also provided in the form of newsletters, cartoons, local radio broadcasts and even videos where residents help plan the script, provide their homes as film locations and act out the scenes themselves. In this way videos have been produced on home maintenance, condensation and energy efficiency, including versions that have been translated into community languages. Urban Care information has also formed part of an interactive television experiment. Programmes on energy efficiency and condensation have been produced with a community group and delivered interactively via a television network. Interactive learning programmes may soon be delivered by cable, microwave or satellite broadcasts into people's own homes. Using digital televisions in this way means that local information programmes can be made and targeted at a specific audience. Soon, low-income families may have increasing access to the educational benefits of new technology without the need to purchase expensive computer hardware or software.

Many householders consider that access to independent advice is an essential prerequisite of spending money on work in the home. They are concerned that only necessary work is undertaken and that any work is completed to a good standard. Unfortunately, most residents find that the only people who are prepared to advise them are builders, and they are far from being independent. The idea of a community group employing their own home maintenance adviser is now working in practice and many groups offer advice as part of complete home maintenance service. The advisers are employed by residents' management committees, which are elected annually from the local community. The role of these community managed Urban Care Projects is to provide information, advice, home surveys, links to builders and training in people's homes on a range of home maintenance topics including heating, energy efficiency, ventilation and condensation damp. The projects can help householders with maintenance planning and, where they have particular problems, help to identify a solution. This means networking with other urban care initiatives to provide specific support and resources.

Projects also provide a direct practical care service where a community worker carries out home maintenance tasks in the homes of older people or residents with a disability. They may undertake essential maintenance work or ensure that the building structure is weatherproof. They will also assist with works that improve safety and thermal comfort within the home. Projects work closely with network installers to ensure that residents benefit from Home Energy Conservation Act funds when they fulfil the criteria for a grant.

15.7 Health

Birmingham Health Authority has recognised areas of health inequality within the city and one of the factors in this inequality relates to people living with poor housing and environmental conditions. The Director of Public Health is quoted as saying *"Men and women who experience adverse social, economic and environmental influences throughout their life do not only tend to die younger but also spend proportionately more of their shorter life span with an illness or disability which limits their function"* [Birmingham Health Authority, 1995].

The health authority has supported the drive to improve people's health through improving their living conditions. Particular support has been provided in the past for projects referred to as "Health-Help" which provided assistance for a range of housing issues with short term project funding. This support has targeted specific client groups, including older people, people with a disability and children under five years. This partnership between the health authority, local authority and community continues within the urban care strategy to provide specific services for vulnerable residents and empower the wider community to act. Mainstream funding is provided by the health authority to co-ordinate health-help activities with home maintenance advisers and practical care projects. Co-ordination is most important, as there is a potential for projects to duplicate activities. Duplication may arise because in practice there are many links between health or home safety work and other works considered as structural maintenance.

A good example of this is cold and damp housing which is more likely to predispose residents to illness or injury. There is considerable evidence that a cold home environment will increase the likelihood of respiratory illness, heart attacks, strokes and accidental injury. In addition, cold and damp conditions in homes can give rise to condensation, fungal growth and increased numbers of house dust mites, which are linked with respiratory conditions such as asthma. A recent urban care survey has indicated that 46% of households in inner city Birmingham suffer with condensation and 43% of these have someone in the household with asthma or a respiratory illness [Birmingham City Council, 1998]

In 1997 it was reported to Birmingham Housing Committee that health service consultation rates for asthma had increased by 191% in the 10 years up to 1993 and that respiratory diseases had become second highest in the list of avoidable hospital admissions, especially with reference to children [Birmingham City Council, 1997]. Since the time of this report, Birmingham City Council and Birmingham Health Authority have collaborated with community groups and other agencies to tackle some of the links between ill-health, poor housing and fuel poverty. Recently this health alliance has co-ordinated the training of community workers from various urban care projects. This has increased the workers' ability to provide information, advice, training and practical assistance in relation to safety, condensation, energy efficiency and fuel poverty. For example, in 1999 the national safety organisation RoSPA organised a unique home safety course for community workers, including training relating to home accidents and cold indoor temperature. Prior to this community workers completed training which earned them the City and Guild Certificate in Energy Awareness. In this way a team of community advisers has been enabled to address issues related to fuel poverty in a cost effective way, even though advice is often provided on a one to one basis in residents' own homes.

In Sparkhill, one of the most deprived inner-city areas of Birmingham, five residents' associations have established a 'community shop' on the high street where volunteers provide advice for local people. It also forms a base for two community managed urban care workers who can provide follow-up services. In Sparkhill and other areas of the city this service includes energy efficiency works for older people and residents with a disability. In addition, the health authority supports workers in providing and fitting home safety aids. Also, the fire service has supported schemes with the provision of smoke alarms and carbon monoxide detectors. Likewise police have supported security schemes which attempt to balance the fear of burglary with the need to adequately ventilate properties. The local authority has supported services which enable or undertake home and environmental maintenance. All of these activities have some impact on health, including stress and related mental illness which debilitates so many inner city residents.

The health authority and local authority are piloting 'Healthy Homes' projects in the Ladywood and Hodge Hill areas of Birmingham. £150,000 of capital was made available by the health authority to pay grants of up to £2,500 for specific works in a patient's home where it could generate a tangible health gain. Patients are nominated by General Practitioners where it is considered that the likelihood of treatment or a hospital admission could be reduced. This criteria includes works to remove cold and damp living conditions in the homes of patients with respiratory illness. The scheme is delivered through home improvement agencies, which can offer a range of housing advisory services in addition to necessary technical support.

15.8 Community training

Another aspect of the urban care strategy is to deliver training to residents to provide them with the skills necessary to recognise what is required within their own homes and to deal with any resulting works. Planning for the maintenance of a home and acquiring some DIY skills both form elements of an urban care training course. The course is often provided to people who have difficulty entertaining the idea of entering into any form of training. Therefore, it is important that the training is delivered on residents' own terms, usually at a local venue. Courses attempt to create a fun and friendly environment in which new ideas and skills don't appear as a threat. The course is divided into modules to allow for a flexible style of learning which meets the aspirations of the participants. Specialist modules are included on energy saving, damp and condensation. The course is accredited by the Open College Network so that credits can be gained towards NVQ levels 1 and 2. South Birmingham College supports the urban care course and draws down funding to pay for community tutors. Residents are supported in training to become tutors to enable them to deliver the course within their own communities. In this way the gradual cascading of information and skills within deprived areas may be far more effective and sustainable than big, expensive campaigns which attempt to change maintenance or energy behaviour.

15.9 Tool loan

Once a resident has gained the skills to undertake DIY works, purchasing all the necessary tools required for a job may be a financial barrier. In some cases community groups have developed tool loan schemes, which make home maintenance equipment and tools available to residents in an agreed area. Schemes are usually adopted by community groups, who recruit volunteers and provide them with the necessary accommodation, training and insurance. The local authority or private sponsorships may provide a start-up grant to purchase tools, after which the loan scheme becomes self-financing through residents donations.

15.10 Community builders' list

All surveys and consultations have highlighted residents' anxieties about finding a reliable builder. The fact that anyone can call themselves a builder, whatever their qualifications or experience, does not give householders the confidence to part with money for maintenance and energy efficiency works. Recent national and local publicity around the issue of 'cowboy' builders has identified problems of unnecessary work being carried out, over-charging for services and sub-standard jobs being completed. This appears to have had the effect of continuing to reduce the amount of housing maintenance work undertaken. In addition, many older people and single parent families are quite concerned about allowing a stranger into their home.

Under the urban care programme residents' groups have established Community Builders Lists. This consists of a record of builders with a range of trades who have a proven record of workmanship, reliability and value for money. The list is owned, controlled and operated by the community groups involved. Six residents' groups initially came together to commission a consultant to establish a vetting system that is workable and legal. Their list is regularly updated in response to the monitoring information collected from the "users' satisfaction questionnaires" which are completed by each resident using a contractor.

Residents can choose an appropriate builder from the list at a number of community locations. There they are also provided with general information about agreeing the works, negotiating a price and avoiding disputes with the contractor. They are encouraged to sign a minor works contract with anyone they employ and a draft copy is provided for this purpose.

Besides the City Council's Housing Department the community builders' list is also supported by the police, trading standards division and Federation of Master Builders. Most importantly it is supported by reputable contractors who want to advertise their services as well as to establish a good reputation for the work they do. The list may also encourage new businesses trying to establish themselves in areas of high unemployment.

The ultimate test is that a list is being used by householders who now have the confidence to employ builders to work on their homes. Community groups have reported very good relationships being developed by groups of residents and particular builders thus avoiding the concerns of having a stranger in the house.

15.11 Resourcing work

With well over 90% of all householders who apply for a house renovation grant failing to get one, financial support within urban care has to be much more than a grant referral exercise. Alternative methods of raising finance have to be considered as it is morally wrong to raise residents' awareness of damage to property, damage to health, risks to safety and security and then offer no solution to these problems for low income householders. Already many inner-city residents are living in fuel poverty which, in practical terms, means that they are faced with a choice of getting into debt or living with low indoor temperatures.

Besides grant advice, urban care initiatives may link to 'benefit take-up campaigns', money advice surgeries, information and training related to financial options and 'save and repair' schemes. Saving schemes may be linked to a community bank or credit union where savings can build up for expenditure on a property and, if necessary, loans made to people who would not be considered creditworthy by the big banks. Some low-income householders may receive assistance from practical care projects while others can save money by doing work themselves on completion of a training course.

Working within a LETS scheme may avoid the need to spend money on employing contractors to maintain a home. LETS stands for Local Exchange Trading System. It enables people to trade services without exchanging money. People form a non-profit making club to exchange services, using their own local "currency" and accounting system. A directory of skills is circulated and the services offered can range from draught-proofing a house to baby-sitting. Members trade with each other whenever they wish, writing out LETS "cheques" to pay for a service. Urban care projects, therefore, network with existing schemes and support the development of new ones.

15.12 Young people

Support for the next generation of householders is given through school packs. Teachers and community education officers have been involved in preparing urban care projects on housing, health, energy and environment, which can be linked to the national curriculum. Teacher seminars have helped review the material and disseminate it across the city. Some schools have gone a step further and started to deliver urban care services within their own community. An example is Yardley's Secondary School which has effectively set up an elected community group made up of pupils from within its own school. They agree an action plan which links up with the plans of the wider community. At a community seminar older people living in the area around the school had said that they had difficulty in tackling some of their large housing problems as they couldn't even change light bulbs when they blew. Young people from the school offered to help by running their own light bulb changing service. They put in a successful bid to the health authority for a grant to buy low energy light bulbs as they last ten times longer than an ordinary bulb and save energy. The health authority accepted the argument that this activity could reduce the incidence of broken bones amongst older people. This school project has since expanded its services within the community and ensured that its activities are sustainable without grant aid. The projects involving young people have been very well received within their community and have boosted the reputation of their school.

15.13 Communication

Once schemes are in place they may be promoted locally by the residents themselves. Volunteers in each street are sought to act as an information link to the neighbourhood. All local schemes provide information on their services, opportunities for further community involvement and a contact point. This information is provided simply on one A4 sheet of paper and photocopied for the "link people" to disseminate. In this way local people take responsibility for the information and it is they who ensure that it gets to the residents who most need to know.

15.14 Achievements

Within the urban care framework 61 initiatives have been established and all have some form of client involvement. In fact 82% are managed by the communities benefiting from the service. Continuous monitoring and evaluation systems are built into initiatives. Regular reviews have revealed the need for improvement in a number of areas, including effective communication, better forward planning and more realistic development timescales. Monitoring data indicates that during the 1998/99 financial year over 18,000 households received urban care information and almost 2,500 households benefited from a community worker visiting their home to provide individual advice. Community training in its various forms involved almost 800 residents. The overall effectiveness of urban care in influencing general trends has not yet been assessed but follow-up surveys to specific area-based initiatives have indicated a significant improvement in home maintenance activity. Residents who undertook works in their home did so with the benefit of improved information, advice and training. Independent monitoring in the

Tyseley Urban Care Area indicated that 35% of owner occupier clients and 50% of council tenant clients would not have undertaken works without the urban care services [Joseph Rowntree Foundation, 1999]. In another area, maintenance expenditure on owner occupied properties was found to increase by a multiple of four in the 12 month period following the launch of community initiatives.

Although it is difficult to assess how much maintenance work is enabled in homes across the whole city, limited follow-up surveys over the 12-month period identified 5,188 completed works. Although there is considerable overlap between the urban care topic areas, a general breakdown of these practical tasks are:

Energy efficiency/fuel poverty	30%
Home safety	23%
General property maintenance	20%
Environmental works	20%
Security improvements	7%

These figures do not include any of the grant aided works or the general activities of partner organisations such as home improvements agencies, Age Concern, Victim Support, credit unions, or local exchange trading schemes.

Although benchmarks are not yet available for urban care projects, services appear to be cost effective. Also the partnership work between agencies and the community indicates the potential for a sustainable service as opposed to short-term projects. In fact sustainability is probably an impossible goal unless ordinary people support the urban care strategy aims to the extent that they are prepared to take action within their own homes or communities. A hopeful sign for the future is that 267 community volunteers already give support to community management or the delivery of the services.

15.15 Conclusion

The urban care strategy has been developing within Birmingham to prevent urban decline in many of its manifestations within homes and neighbourhoods. It aims to break away from limited projects with short-term solutions, which can encourage a dependency culture. Instead its strength is in the empowerment of individuals and communities to take action.

The partnership approach is an essential element, not only because it enables more effective services to be developed, but also because it ensures that those community services are sustainable using affordable resources. Partner organisations are now building urban care into their own plans. Birmingham Health Improvement Programme, which has been produced jointly by the health authority and local authority, proposes an expansion of urban care across the city to tackle unhealthy homes. It particularly points to the need to improve housing conditions related to the increase in asthma and respiratory illness [Birmingham Health Authority, 1999].

The incidence of cold, damp homes and related fuel poverty is inextricably linked to many other factors that lead to urban decline. Urban care has tried to link these various elements within a comprehensive strategy. The resulting local action plans may appear as a jigsaw of activities, which provide information, advice, skills, training, resources and practical assistance. However, given a degree of co-ordination, all these activities are complementary and benefit greatly from partnership working and community management.

16

Working in partnership: lessons from ten case studies

Emma Jones, Victoria Wiltshire and Helena Poldervaart

Partnerships between health and local authorities for practical action to improve health by increasing the energy efficiency of housing are relatively uncommon. This chapter presents the findings from a research project looking at ten such partnerships. It begins by explaining the aims and methodology of the research before presenting a summary of the ten schemes. The barriers to multi-agency working are then discussed with suggestions for how these might be overcome. The chapter ends by recommending support that could be provided at both a national and a local level to facilitate multi-agency working, with suggestions for further work that is required in this area.

16.1 Introduction

The link between cold/damp homes and health is now recognised by the government [DoH, 1998] and there is considerable literature examining this relationship [Henwood, 1997; ACE and PIP, 1999].

Practical action to improve health by increasing the energy efficiency of housing requires a partnership approach. As providers of healthcare, health authorities and/or primary care groups should be involved as well as housing professionals within local authorities. The importance of partnership working is recognised by both the Department of Health (DoH) and the Department for the Environment, Transport and the Regions (DETR) in recent Green (consultation) and White (Government policy) papers [e.g. DoH, 1998b; DETR, 1998]. Michael Meacher, Minister of State for the Environment, recently wrote, *"We are coming to realise more and more how people's health is affected by a wide range of factors - economic, social and environmental - and that to make real and lasting improvements we need to tackle problems across a broad front. This requires new ways of thinking and new partnerships at all levels... ".* A full discussion of partnership working in this area can be found in ACE and PIP, [1999].

Included among the partnerships examined in this chapter are the three case studies presented earlier in this book (in Birmingham, Cornwall and Nottingham).

16.2 Research

Research was carried out by ACE and PIP to investigate barriers facing multi-agency approaches to housing and health. The research was funded by Transco and Eaga Charitable Trust (Eaga-CT), with additional contributions from EAGA Ltd, Unison, The Energy Saving Trust (EST) and the Local Government Association (LGA). The objective of the research was to disseminate best practice to local authorities and health authorities whilst also making recommendations for facilitating multi-agency work.

A number of case studies were chosen based on evidence of a partnership approach between the local authority and the health authority. For each scheme, interviews were held with all the relevant partners, often in a group discussion followed up by individual telephone interviews. Interviews focused on details of the scheme in question, lessons learned from the approach taken and views from the officers involved about the success of the scheme.

The second stage of the research involved holding workshops with key partners in three of the schemes, to consider with them the barriers they have encountered and how these might be overcome. A national working seminar was then held, attended by representatives from some of the schemes, members of the advisory group and other key national players. Following an initial presentation of the work carried out, the seminar was run as a workshop in order to develop recommendations for facilitating multi-agency working.

The final part of the project will involve dissemination of the results. In autumn 1999, all local authorities, health authorities and primary care groups will be sent a copy of a 'Step-by-Step Guide' to setting up partnership initiatives. In addition, the full report and case studies will be published.

16.3 Case studies

The ten schemes varied in approach but all involved a partnership between the local authority and health authority or NHS Trusts. Some also involved the local Energy Efficiency Advice Centre (EEAC), the local utility and/or voluntary agencies.

The majority of schemes were concerned with general health issues arising from cold and damp homes, but three specifically targeted asthma. Seven involved training health visitors while three resulted in the set up of referral systems with GPs. Six had funding directly from the health authority or through joint finance and five had funding from the DETR, via the EST-administered 'HECAction' programme. (HECAction offers pump-priming funding to local authorities for schemes to encourage energy efficiency.) Only two had secured ongoing funding, with six attempting to find additional funds, while two had ceased.

A. HECACTION FOR IMPROVED HEALTH AND HOUSING

Telford and Wrekin Council, Shropshire Health Authority and Telford EEAC are working together on this project. It was funded by a £50,000 HECAction grant plus £13,000 from joint finance and £25,000 from Council's Home Repairs Assistance Grants (HRA). The scheme commenced in 1997 and is ongoing. 35 health workers have been trained and grant-funded energy efficiency measures have been installed in the homes of around 50 patients referred by their health visitors. Exhibitions were also held in GP surgeries on the link between health and energy efficiency.

B. HEALTH AND ENERGY COMBINED ACTION

This scheme is a partnership between Eastleigh District Council, Solent EEAC, Southampton Community Health Services NHS Trust and the Winchester and Eastleigh Healthcare NHS Trust. It was funded by a £53,000 HECAction grant, plus some financial support form Eastleigh DC. The scheme commenced in 1997 and is ongoing, although activity is now dwindling. It involved providing information to health visitors on energy efficiency advice provision (there are no figures on how many were involved) plus the offer of subsidised energy efficiency measures to those on benefit recommended by their health visitor (of which there has been minimal take-up).

C. ENERGY ACTION FOR HEALTH

In this scheme, Doncaster Metropolitan Borough Council is working in partnership with Doncaster NHS Healthcare Trust.   Funding was provided from joint finance plus a small amount from HECAction, with the Council's HRA Grants used to fund measures. Health workers were trained to provide energy efficiency advice and information on grants. Home Energy Efficiency Scheme referrals in the region were increased by 15% as a result, and 45 houses had measures installed, funded by HRA grants.   Energy efficiency advice surgeries were also conducted in GP clinics

D. ENERGETIC HOMES

This scheme is a partnership between Leicester City Council, Leicestershire Health Authority and local GP practices. It was funded by a HECAction grant and by European funding. It started in 1997 and is ongoing. The scheme involves a GP referral system for the provision of grants for heat recovery systems in the houses of asthmatic patients; 40 households have had measures installed.

E. ENERGY EFFICIENCY TRAINING FOR PRIMARY CARE WORKERS

Hammersmith and Fulham Council worked with the Riverside Community Health Care Trust on this HECAction funded scheme which ran from 1996 to 1997.   Energy efficiency advice training was provided for primary care workers, plus a range of reference booklets for the workers and their clients and the production of a simple to use modular training programme; 74 people were trained as a result of the scheme.

F. BEAT THE COLD

This scheme involves Beat the Cold working in partnership with North Staffordshire Health Authority, The Work Experience Centre, NEA Birmingham, British Gas and Midlands Electricity. It has been funded by a variety of sources, including the health authority, British Gas, NEA, and Midlands Electricity. The scheme started in its present form in 1996 and is in the process of becoming a charity. Carers and community/voluntary workers such as Age Concern staff have been trained, and an awareness campaign amongst the elderly has been conducted. Around 50 households have had measures installed.

This scheme was chosen for a workshop. Participants included members of the health and local authorities plus local voluntary organisations, utility and insulation company representatives.

G. 'SNUG' SCHEME

In the 'SNUG' scheme, Birmingham City Council's Urban Renewal division is working in partnership with the Family Health Service and Birmingham Health Authority. £150,000 of funding was provided by the health authority for the scheme, which started in 1996 and is just coming to an end. It involves a GP referral scheme for the elderly to receive grants. To date, 45 GPs have referred 120 householders. This scheme was chosen for a workshop, attended by representatives from the council, the health authority and community groups. (Initiatives in Birmingham are discussed in detail in Chapter 15.)

H. ENERGY, HOUSING AND HEALTH

The Energy, Housing and Health scheme (*see* Chapter 13) involves Nottingham Health Authority working with Broxtowe, Rushcliffe, Gedling and Ashfield District Councils, Nottingham City Council, Nottinghamshire County Council and NEA Nottingham. £8,500 was provided from the Health Action Group (see Chapter 16) Health Initiatives Budget plus funding from Eaga-CT, while Nottingham City Council funded new central heating units.

It started in 1996 and is ongoing. The scheme has involved energy efficiency measures and heat recovery ventilation units being provided to low income tenants of the council with asthmatic children. 7 households have been involved. In addition, training has been provided for health and housing professionals (no figures are available on how many).

I. CORNWALL AND ISLES OF SCILLY

This scheme has been led by Cornwall and Isles of Scilly Health Authority working in partnership with GPs, schools, local authorities and the local EEAC. £300,000 was provided from the Health Authority. The scheme operated from 1995 to 1997 and monitoring has recently been completed (and is described more fully in Chapter 12). It has involved central heating being installed in selected homes with asthmatic children; 101 homes have been improved as a result of the scheme. This scheme was chosen for a workshop.

J. OXFORD CITY COUNCIL

The 'medically supported heating scheme' is funded by Oxford City Council. It makes use of a number of referral networks, including GPs and health visitors, to identify priority cases for insulation and heating improvements. It has been in operation since 1987 and initially had health as well as local authority funding. Over 3,000 homes have been improved under the scheme.

16.4 Multi-agency working: Barriers and opportunities

The research on which this chapter is based was concerned with identifying barriers to multi-agency working. Every scheme studied had achieved some success and all those interviewed could identify benefits of multi-agency working. However, all have also encountered barriers, which for some schemes have significantly limited their results. In this section, the findings regarding benefits of and barriers to multi-agency working are presented. This is followed by suggestions for how these barriers could be overcome.

BENEFITS OF MULTI-AGENCY WORKING

Numerous benefits were cited by workshop participants. It was agreed that multi-agency working can help to improve levels of service delivery by facilitating the adoption of best practice. In particular, a wider understanding of issues develops when different agencies are involved and both information and joint problems can be shared. Professional barriers and other boundaries can be broken down through working with other agencies and sharing resources and knowledge can lead to increased efficiency of working practices.

BARRIERS TO MULTI-AGENCY WORKING

Securing ongoing, longer term funding is a problem experienced by a number of schemes, and this can be related to other issues such as incomplete research evidence and monitoring difficulties. Ownership of budgets can also be a problem. Several reported that structural barriers to cross-sectoral funding still exist. Multi-agency working can also result in a strain on staffing resources due to the time required to develop and maintain the partnership.

There is still no generally accepted robust piece of research that proves the link between housing and health and quantifies the benefit of investing in energy efficiency. Health authorities in particular may require quantitative evidence on the cost effectiveness of a new approach before they will commit money, or in some cases time. While a number of research projects are in the process of being conducted, it is unclear whether their methodology will be considered to be sufficiently robust by health professionals.

Different agencies have different objectives which can result in conflicting priorities and timescales. Some projects felt that health authority priorities are focused on service provision rather than prevention of ill-health while the local authority is concerned with increasing energy efficiency and providing access to affordable warmth.

Inflexibility in organisations was cited as another barrier and multi-agency working sometimes leads to 'buck-passing' because no-one wants to take responsibility.

It was reported that 'professional snobbery' is sometimes encountered and, in particular, health professionals are often unwilling to commit to a local authority-led project. Political interplay between agencies can obstruct effective working. At times these difficulties lead to actual conflict between partners. Partnerships frequently have an unequal power balance, depending on the type of project or source of funding.

Limited history of working together is a barrier mentioned by some of the case studies. In general, the longer the partnership has been in place, the better the partnership will work. This lack of experience of joint working can mean that the person planning the scheme does not have the right contacts in the other key organisations, and therefore may not involve the most suitable people from the outset. This in turn can result in those people lacking a sense of ownership of, and therefore commitment to, the project.

Good communication is essential for effective multi-agency working – in particular, ensuring that the right people receive appropriate information. Contact between agencies often exists at a front-line level but not at a strategic, policy level.

In addition, there are practical and perceived problems with free exchange of information in relation to the Data Protection Act and patient/client confidentiality.

HOW BARRIERS COULD BE OVERCOME

Joint working should be recognised as a key function and consideration might be given to this by allocating budgets to it, in the same way that, for example, monitoring is budgeted.   Longer-term funding would reduce the amount of time that has to be expended in seeking funders for projects.   Greater flexibility and openness is required on the part of some funders, as well as a move away from an emphasis on pilot projects. A great deal of research into the area of health and energy efficiency has been conducted, but most has not been disseminated widely. A document summarising the relevant research in a format which could be used by practitioners would be helpful.

Guidance from central government on the importance of this kind of approach would also be helpful.   While health authorities often state that they need quantitative evidence in order to commit resources in this area, the Department of Health is convinced of the benefits of partnership working in this area [DoH, 1999].   Further guidance to this effect would be helpful.

It was suggested that a third party could take responsibility for identifying joint working opportunities between agencies.   This could be complemented by an increased use of secondment to move skills to where they are most needed and to break down professional barriers. A further step would be to have some posts jointly funded.

Lack of experience of joint working is a barrier that only time can overcome.   However, the process can be assisted by the setting up of formal partnership agreements at the start of a project.   This would set out the agreed aims of the project, how the agencies intend to work together and what the financial arrangements are.   Such a process builds trust and understanding.   To strengthen this agreement, internal commitment in each organisation should be obtained, particularly at departmental level, ensuring that the project agenda is built into the business/work plan.

Both health and local authorities hold data which are key to targeting resources. These can and should be shared more freely to mutual benefit.   However, case studies have demonstrated widespread ignorance regarding what is available and what can legally be shared.

16.5 Recommendations

A number of changes could be made at a local and national level to facilitate multi-agency working.   Third party organisations could be set up at a local level with a responsibility for identifying joint working opportunities.   Their duties could include liaising with all relevant agencies, determining the relevant contacts and identifying how the differing priorities could be achieved through joint working.   This role could be taken by the local voluntary services council. A less resource intensive option would be to set up a cross agency networking structure at a local level.

On a national level, guidance from central government departments on the desirability of multi-agency approaches would help to overcome many of the barriers. In particular, giving local agencies more control over determining what areas to spend their budgets on would increase the opportunities for these kinds of projects to get funding. It was suggested that an ideal solution would be the creation of ring-fenced budgets to pay for staff time spent in the set up and development of multi-agency working. Funders could themselves assist in building partnerships in practical ways; both by including partnership working as a requirement for funding (as is already the case with Single Regeneration Budget funding) and by actively facilitating partnership development where agencies are not used to this way of working.

Joint working between relevant agencies at a national level is needed to develop best practice guides and systems for their dissemination including seminars. A mechanism is needed to link local experience and research to national policy development, thus enabling a strategic overview which moves away from the present, ad hoc, uncoordinated approach.

16.6 Conclusions

It is encouraging that the issue of multi-agency working, in particular in relation to energy efficiency and health, is receiving so much attention at a national policy level. Both the Department of Health and DETR have taken an advisory role in the ACE/PIP project and are already meeting cross-departmentally.

Some of the ideas emerging from the case studies would need a high level of such co-operation to be developed, but we believe the time is right to take them forward. The Health Action Zones are specifically intended to encourage and enable multi-agency working and could be used to pilot some of the recommendations. Similarly the Energy Efficiency Partnership, a multi-agency partnership facilitated by the Energy Saving Trust, is already starting to initiate and develop mechanisms for delivering energy efficiency measures through health professionals such as GPs and health visitors, under the new Home Energy Efficiency Scheme. These measures are targeted at the fuel poor and thus will achieve the greatest health gain.

It was striking, throughout this project, how many practitioners and researchers were found to be working in comparative isolation. The benefits accrued when they were able to meet and exchange ideas proved to be significant. Dissemination by the printed word is important but our experience demonstrates the immense value of face-to-face contact. A pro-active approach is required, to enable this kind of cross-fertilisation of ideas and experience and to co-ordinate effort so that the same research is not duplicated, while other important areas are neglected.

Part Four:

Ways forward

17
The cost of poor housing - and how to reduce it

18
Healthier homes: the role of health authorities

19
The unavoidable imperative: cutting the cost of cold

17

The cost of poor housing - and how to reduce it

Peter Ambrose

17.1 The origins of the research

The set of ideas and the research reported on in this chapter had their origins in a well-heated Transit Lounge at Frankfurt Airport. In 1992 the author was appointed by the Foreign and Commonwealth Office 'Know How' Fund to select and lead a team of British housing experts to advise the Bulgarian Government on how best to reform their housing policies in the situation following the ending of Socialist conditions.

The advisory team that emerged, some of whom played a very active part in the research, included senior figures from the Housing Corporation and the Housing Finance Corporation, an estate regeneration specialist, an architect and a lawyer in private practice, a retired Chief Planning Officer, a local authority housing manager, the Director of a housing association and two academics - Alan Murie and the author. The origins are important because it is rare for a team with this broad combination of expertise to collaborate on a research project. The subsequent history of the research reflects the mix of professional competencies, the combination of private, public and voluntary sector perspectives and the inter-play of practitioner and academic approaches that characterised the work of the team in Bulgaria.

It might be noted that in that country the team encountered problems related to the poor quality of construction and maintenance of much of the high rise stock, heating systems with rusted-up regulator valves, deeply non-participatory traditions of management, a conspicuous lack of care for all public areas in the housing schemes, rapidly rising rents following de-restriction and, by 'western' standards, a high degree of overcrowding. What they did *not* come across was much evidence of 'street homelessness'. And when they sought to convey that in prosperous capitalist Britain there were about 30,000 or so 'excess deaths' most winters as a result of poor heating and insulation they were politely disbelieved and invited to yet another glass of Bulgarian red.

17.2 The CEHI research programme

In 1993, on the return journey from the last of the seven visits made to Bulgaria, the team agreed that they had done all they could to convey the lessons of 150 years of British housing policy - good and bad. Wishing to continue the collaboration in a more familiar context they determined to initiate a research programme that would explore what was perceived to be one of the great gaps in our understanding - the nature, extent and monetary value of the 'cross-sectoral' costs to other services generated by poor housing conditions.

The result was the Cost-effectiveness in Housing Investment (CEHI) research programme based initially at the Centre for Urban and Regional Research at the University of Sussex and now in the Health and Social Policy Research Centre at the University of Brighton. The Programme is overseen by a Management Committee which includes most of the Bulgarian team. The Steering Group has been chaired since its inception by Stephen Hill of Capital Action Ltd and the Programme is directed by the author.

The aims of the CEHI programme were agreed as follows:
- to show that investment in better quality housing will produce more than commensurate reductions in 'cross-sectoral' costs (costs falling on budgets other than housing);
- to identify, systematise and where possible evaluate these cost savings;
- to identify what forms of additional investment in housing quality will be most cost-effective;
- to promote a more informed debate at all levels on these issues.

A wide range of public, private and voluntary organisations and several professional institutions, notably the Royal Institution of Chartered Surveyors, showed immediate interest in these aims and funding to initiate the project was quickly raised from the Institution and other sources.

An early task was to define precisely what was meant by 'better quality housing' and after long discussion a fully worked out definition was arrived at [Ambrose, 1996]. A brief and cross-culturally acceptable version of this definition was found in a work by Seedhouse [1986]:

'A satisfactory housing standard is one that provides a foundation for, rather than being a barrier to, good physical and mental health, personal development and the fulfilment of life objectives.'

17.3 Poor quality housing and 'exported costs'

In recent years considerable research has focused on the interface between housing quality and health status. The interface is a complex one and it is futile to search for simple 'cause/effect' relationships. But evidence gathered from many studies shows clear patterns of association between poor physical conditions, for example cold, damp, infestation and overcrowding, and an increased incidence of ill health. Equally living in a stigmatised area with poor services, high benefits dependency and a high level of fear of crime was almost inevitably stressful and was likely to impact on mental health. A comprehensive collection of essays on the housing/health connection has been published [Burridge and Ormandy, 1993]. It is obvious that a growing incidence of ill-health must add to costs for health services which are already under increasing strain in Britain and other European Union countries as a result of various factors including ageing populations [see also other pioneering work on the costs issue by Boardman, 1991, Carr-Hill *et al.*, 1993 and Lawson, 1997].

But it was evident to the CEHI team, whose members had had very considerable experience in a number of housing fields, that the connection between poor conditions and high costs was multi-dimensional. Poor living conditions appear to be generating additional costs not only to health services but also to other key service providers. These include:

- the education service (because children living in cold, damp and overcrowded home conditions cannot learn as effectively)
- the police and judicial services (because poor housing design and construction is associated with a higher incidence of some crimes)
- the emergency services (because poor housing conditions and reliance on 'secondary heating' increase accident and fire risks)
- the energy supply services (because poorly designed housing uses excess energy and produces ecological damage).

Over three hundred research studies examining these issues were reviewed as an early part of the CEHI programme of work [Ambrose, Barlow, *et al.*, 1997]. Although few of these studies explicitly addressed cost issues it seemed evident from many of them that poor living conditions implied additional costs for a wide range of service providers. The CEHI team termed these costs *'exported costs'* because they are generated by under-investment in one sector (housing in this case) and then 'exported' to others.

17.4 A matrix of 'exported costs'

The team began to list and systematise the range of housing-associated costs that might well arise. At least four categorisation axes were apparent:

1.   Capital Costs *versus* Revenue Costs

2.   Costs felt on the personal finances of individual residents *versus* costs felt 'externally' by service providers of one kind or another (although some of the latter no doubt work through to the individual in the form of higher taxes)

3.   Costs that can be thought of as 'systemic' in that they impact regularly, and sometimes imperceptibly, as life is lived *versus* costs felt in more visible and 'formalised' ways (such as the annual bid for funds by a public service whose funding formula recognises the high cost of service delivery in run-down areas or the costs of special response programmes such as New Deal for Communities).

4.   Categorisation in terms of the susceptibility of each cost to accurate measurement. The categories adopted here are:

H	'hard' costs that can be precisely quantified
M	costs of 'medium' quantifiability
NQ	costs that clearly exist but are currently non-quantifiable

Working with this four-fold categorisation scheme the team developed the following matrix:

A matrix of costs whose levels can be related to poor housing			
	COSTS TO RESIDENTS		**COSTS TO OTHERS**	
ystemic Capital	high annual loss of asset value if property rented	H	high annual loss of asset if property owned	H
ystemic Revenue	poor physical health poor mental health	H/M M/NQ	higher Health Service costs	H/NQ
	social isolation	NQ	higher care services costs	M
	high home fuel bills	H	high building heating costs	H
	high insurance premiums uninsured contents losses	H M/NQ	high insurance payments	H
	spending on security devices	H	spending on building security	H
	living with repairs needed	NQ	high housing maintenance costs	H
	under-achievement at school	NQ	extra costs on school budgets homework classes at school	H H
	loss of future earnings	M	loss of talents to society	NQ
	personal insecurity	NQ	high policing costs	H/M
	more accidents	M	high emergency services costs	H
	poor 'hygienic' conditions	NQ	high Environmental Health costs	H
	costs of moving	M	disruption to service providers	M
	adopting self-harming habits	M	special health-care responses	H
ormalised Capital			Government and EU programmes, SRB, New Deal, etc	H
ormalised Revenue			'Statements of need'	H
			Section 42	H
			HIP statements	H
			Police funding formula	H
			Fire and Ambulance services funding formulae	H

[Ambrose and Randles, 1999]

A matrix of this nature not only provides a theoretical framework for the task of estimating some of the total costs incurred but it also prompts a number of questions requiring further examination:

- how is the cost of bad housing distributed between residents and servicing agencies?
- of those felt by the latter, which agencies bear most costs?

- which agencies might therefore save most as a result of increased investment in housing?
- which costs are currently poorly recorded or measured?
- how do revenue costs and capital costs compare in terms of 'weight'?
- what *forms* of increased investment in better housing might most reduce 'costs-in-use' and capital depreciation costs?

By identifying a range of more measurable costs (H), the matrix also gives some guidance concerning the most promising ways to continue the task of evaluating 'exported costs'.

17.5 Case studies and empirical findings

IN CENTRAL STEPNEY

Early in 1995, in view of the progress made, the CEHI team was commissioned by the London Borough of Tower Hamlets to carry out a 'Health Gain' study to compare the health status of a population before and after a major housing improvement. The housing to be improved was on two estates in central Stepney, the Limehouse Fields and Ocean estates, then some of the worst housing in London.

The housing renewal was part of the Central Stepney Single Regeneration Budget (SRB) improvement programme. The introduction of the SRB policy was a shift of ethos and represented the first opportunity to plan investment strategies to produce benefits which cross departmental boundaries [Hill and Barlow, 1995]. This placed a greater premium on research, such as the CEHI programme, which sought to evaluate the costs of *not* working in a cross-departmental and holistic manner. Apart from the vital importance of evaluating some of the benefits of housing improvement for the housing authority the project came at an opportune time to subject the central propositions of the CEHI programme to some empirical test.

The 'before' element in the study was carried out over the winter of 1995/6. A total of 107 households (525 residents) was interviewed using an intensive survey methodology. The technique was to collect data on the health of all household members using unstructured interviews and several call-backs per household over a five month period. The work was carried out by bi-lingual pairs of interviewers since 83% of the population spoke Sylheti as a mother tongue. The response rate was about 95%. The 'after' survey, on the same households following re-housing in improved conditions, will be carried out in the winter of 1999/2000.

The housing conditions encountered in the 'before' survey were extremely bad [Ambrose, 1996]. Over 47% of the rooms were damp and 69% of the population reported that the heating did not keep them warm enough. Over one third of households suffered from infestation from cockroaches and pharoah ants and the room density was well over the legal limit at 1.43 people per room. The 107 households reported 280 Illness Episodes over the survey period and there was a total of 29,114 Illness Days, about 37% of the total person/days. The main ill-health suffered was coughs and colds, aches and pains, asthma and bronchial problems, digestive disorders and depression.

The relationships between on the one hand (a) dampness, (b) lack of warmth and (c) accommodation needing repairs and, on the other, the incidence of coughs and colds were all significant at the 99% level. Damp households and cold households experienced over twice the rate of Illness Episodes than dry and warm households. Residents themselves, when specifically asked about the matter, overwhelmingly regarded Illness Episodes as 'very closely related' to housing conditions and especially to poor, unreliable and expensive heating systems and damp penetration.

This evidence of a link between poor housing conditions and poor health gained from residents was fully substantiated by a second survey - a round of interviews with over fifty providers of health, education, law and order and other services to the central Stepney population. Almost without exception these professionals working in the area also considered that poor housing conditions, especially in relation to cold, damp, poor repairs records and excessive noise, greatly increased the call on their services and/or reduced their capacity to deliver as good a service as they wished. What they were not yet in a position to do was to give an accurate picture of by just how much their costs were increased for these reasons.

These two surveys, one of residents and one of service providers, enabled the team to conclude that some very direct linkages existed between poor housing and a number of adverse outcomes. But they also identified a number of 'indirect' processes that worked to compound the problem and further reduce the health status of populations in very poor and stress-laden environments. These included:

- Lowered resistance to illness, and longer recovery times, related to long exposure to poor conditions
- Adoption of health-threatening habits such as smoking as a means of coping with the stressful conditions rather than as a chosen life-style.
- Reduced self-organising capacity, for example in accessing health and other services and complying with courses of treatment, related to the continual stress of living in a poor environment.

- Diversion of specialist professional expertise (for example the time spent by doctors in writing 'housing letters' or by teachers in giving 'social work' support).

PADDINGTON - A COMPARATOR AREA

It was found that the increased use of primary care and hospital services in central Stepney appeared to be adding substantially to National Health Service (NHS) costs compared to those generated in a comparator area of improved housing in Paddington (an inner urban area of west London). Here, using identical survey methodology over almost the same period, but on a smaller sample, the reported illness rate was about one seventh that in Stepney. An exploratory assessment of the differences in costs generated (in primary and hospital care only) indicates that the annual healthcare costs per household were £515 in the Stepney sample and £72 in the Paddington sample [Barrow and Bachan, 1997]. At least part of the difference can be regarded as costs *exported* from the housing sector to the NHS.

WORK IN HOLLY STREET HACKNEY

On completing the first phase of the central Stepney 'Health Gain' project members of the CEHI team and Capital Action Ltd were engaged by the London Borough of Hackney to carry out further research on the Holly Street redevelopment. The intention was to build on previous research in the area [for example Woodin Consultancies, 1996, Wadhams, 1998 and Whitehead, 1998] and especially to gain a better understanding of the ways in which cost savings in the delivery of the present standard of services, or better service delivery for the same costs, might be achieved by better standards in inter-agency working between the main service providing agencies.

The research aimed:
- to evaluate the findings of the Woodin Consultancies [1996] report;
- to translate the benefits observed, where appropriate, into quantitative indicators with monetary values and/or qualitative scores;
- to develop a greater insight into the meaning of the indicators in order to provide practical assistance to all stakeholders involved in the planning, implementation and evaluation of regeneration programmes, with particular reference to the current stage of the Holly Street project.

In attempting to quantify more of the financial benefits accruing from area improvement the project aimed to give greater validity to the

Option Appraisal process that had long been one of the key procedures used by the Department of the Environment and its successor Department of the Environment, Transport and the Regions when evaluating the likely benefits of alternative investment strategies.

Equally, and in relation to 'quality of life' issues, there was a need to examine more closely the complex ways in which increased safety, health benefits, better educational and training opportunities and similar positive outcomes had mutually reinforcing effects in residents' lives. There was a need to establish what still remained to be done to transform the area into one that people saw as a long-term home and which had its own in-built sustainability. In addition, more needed to be understood about ways of addressing what many residents saw as the isolation of the area and its disconnection from the 'mainstream' of economic life.

Building on the research methodology used in the Stepney work, three main research methods were used to explore various aspects of these issues - a Survey of Residents, a Survey of Service Providers, and a number of Focus Groups. The work was carried out during 1998 and the main findings are detailed in Ambrose and Randles [1999].

The team considered that considerable progress was made in relation to Objectives 1 and 3. In relation to Objective 2, the evaluation of the cost effects, it became evident that far more research time would have to be invested than was available under the project funding if the health benefits of urban regeneration, which had already been reported in largely qualitative terms both in Stepney and Holly Street, were to be quantified. The team concluded that it was vital that this investment in research be made in the near future since an evaluation of the apparently very large savings that could be made by investment in better housing would be valuable evidence to bring to bear in the political arena as national resource allocation issues are addressed.

17.6 Steps to 'design out' high service cost outcomes

Reflection on all the empirical work carried out by the CEHI programme in various locations over the past six years suggests a number of ways in which negative health outcomes and other generators of excessive 'exported costs' can be to a degree 'designed out' over time. Some of the key steps are:

ENSURE BETTER HEATING AND INSULATION

Since much of the rest of this book focuses on this issue the arguments
need not be rehearsed here. British standards in relation to heating and
insulation, or at least the situations they permit, lag behind those of almost
all other European countries with similar climatic characteristics. This is
one factor behind the phenomenon of 'excess winter deaths' which so
surprised our Bulgarian hosts (as compared to the Swedish experience of
a slight excess of deaths in summer). It also helps to explain the appalling
standards of health revealed in the Stepney work [Ambrose, 1996] and the
sharply lower call on health services following housing improvement
observed in Paddington (where improved standards of heating and
insulation were mentioned as a health factor by almost all respondents).

It is important also not to lose sight of the significant cost savings
that arise for the fire protection and ambulance services if the rate of
accidental fires is reduced. In the central Stepney area there was
considerable dependence on secondary forms of heating as a result of the
unreliability and high cost of the main heating systems. This may well be
one reason why the incidence of fires in the home in an area of poor
housing can exceed that in a more prosperous area by a factor of four or
more.

Given indications of this nature it is evident that heating and
insulation standards are in urgent need of revision not just on 'social'
grounds but in terms of cost-effectiveness and economic efficiency.

PROGRESSIVELY REDUCE 'SPATIAL FRACTURING'

Past housing finance, design, construction and allocation policies have
reflected the deeply divisive nature of British social structure over the past
two centuries. The policies have left a significant proportion of the
population in a 'spatially fractured' pattern, that is to say living in areas
that have highly specific tenure and socio-economic characteristics. This
is sharply different from the more socially mixed residential and tenure
structures to be observed in, for example, many Swedish or Dutch
suburban areas.

The result of these social cleavages is that 'area effects' bestow
on some populations a trans-generational and self-reinforcing set of
negative effects – low self esteem and aspirations, poor employment
chances, credit and insurance difficulties, 'labelling' problems, difficulties
in recruiting, for example, teachers, and so on. These life difficulties are
almost inevitably damaging to health. Concomitant adverse effects can be
seen on the morale and even health of those working in public services in

the most stressed areas. This may understandably affect the way they deal with residents, sometimes exacerbating the difficulties of living in the area.

Planning, construction and allocation strategies with the aim, inevitably a long term one, of reducing the degree of spatial fracturing by the explicit recognition of the benefits of 'mixed communities' are likely to carry positive results in terms of better social integration [Page, 1993] and service cost reductions.

FACILITATE BETTER NUTRITION

National housing design norms and construction standards have not been sufficiently mindful of the need to promote and support a satisfactory level of nutrition. Standards need to be re-considered to ensure, *inter alia*:

- reliable supplies of water of a completely satisfactory quality;
- food storage space, both within the housing unit and communally, that allows savings on food costs by bulk buying without the attendant risks of increased levels of infestation;
- the encouragement of some home production of food by designing in some composting facilities and cultivable areas into housing schemes;
- aiding environmental sustainability by designing housing schemes so that rainfall on 'hard surfaces' is conserved for plant watering;
- ensuring that advice on nutrition, food storage, etc. is readily available.

In the context of a book on affordable warmth it is worth pointing out that the lower the proportion of income that needs to be spent keeping adequately warm the more there is available for the other main prerequisite for good health, proper nutrition.

SENSITIVE NEIGHBOURHOOD AND HOUSING MANAGEMENT

More sensitive neighbourhood management arrangements need to be devised including systems for 'trouble shooting' in the form of local 'one-stop' individuals or small teams of advisers who have known office and emergency hours, an easily remembered telephone number with no answer-phones, and operators who can redirect calls to *any* service relevant to such 'multi-agency' problems as may arise.

The aim is to ensure that residents do not have to spend time, money and mental energy on seeking to access a number of agencies themselves but can see 'problems being sorted out in front of them' (as one resident put it in the Holly Street case study). It is crucial to ensure that such advisers are genuinely enabling and that one of their aims is to work towards empowering people to sort out their own problems in accessing services and rights.

In the context of 'affordable warmth' it is worth noting that the high standards of 'hands on' housing management found in the new Holly Street included careful instruction in the proper operation of the heating system controls and continuing help should that be required.

COMMUNITY RESOURCE CENTRES

There is a need to provide community focal points in the form of Community Resource Centres that are *used* because they incorporate many of the services needed to live a reasonably comfortable life. Such Centres might well combine the local medical practice with recreational facilities, 'one-stop' trouble shooters, bulk food storage capability, advice and information on a range of matters including nutrition, cultivation, heating, home maintenance, getting jobs, the local labour market, welfare rights and benefits. The Community Resource Centre might also run short courses on any or all of the above; it might also fulfill such other functions, within practicable limits, as users might identify and wish to organise.

SMALL-SCALE VOLUNTARY ACTIVITIES

There is considerable potential benefit in promoting and facilitating small-scale and organic activities such as befriending schemes, food schemes, credit unions, and religious organisations (on the model of, for example, the Frontline Church in the Toxteth area of Liverpool which runs a weekly Kids Club and visiting programme for about 700 children). Such activities are potentially powerful factors in the thickening up of networks of interaction. They have a capacity to reduce the incidence of ill-health, especially in terms of mental stress, and in some situations can have a therapeutic effect and reduce the need for expenditure on a medical 'cure'.

'DESIGN OUT' CRIME

In numerous surveys, including the work in central Stepney, Paddington and Holly Street, residents have emphasised the fear of break-ins and street crime as one of the main sources of mental stress and sometimes of course physical injury. Much has been learned in recent years about how to minimise the incidence of crime and anti-social behaviour by careful attention to the design of public spaces, improvements to domestic security systems and the use of CCTV. While Crime Prevention Officers are now a standard part of police teams there are some indications that their advice is not yet sufficiently taken account of, especially by the developers of 'social housing' who are operating to tight cost guidelines. It is vital to the cost-effective use of resources that sufficient initial investment is made in crime prevention in order to achieve subsequent reductions in NHS and other costs.

17.7 Conclusions

Much recent research, including the various projects carried out since 1993 by the CEHI team, has shown that the interface between housing standards, health status and other outcomes is a complex one. Certainly fuel poverty is a vital aspect of the problem. 'Affordable warmth for healthier homes' will help immensely and must be pursued vigorously as an objective. But this is perhaps a necessary rather than a sufficient condition to the achievement of higher health standards. The search for better health, and the large-scale cost reductions this should generate, needs to be carried out simultaneously on a number of other fronts. This chapter has sought to identify some of the other aspects of inadequate housing standards that appear to be generating additional costs on a range of public, private and voluntary sector budgets.

Above all, in directing attention to the potentially massive cost savings that should flow from investment in better quality housing, it seeks to re-locate the problem onto grounds that might be more promising in terms of producing progressive political interventions. The issue is, in essence, not a 'social' one. It is about the pay-off to be gained from a more *preventative* approach to health outcomes by paying more attention to the improvement of living conditions. It is, in the end, about improving cost-effectiveness in the use of public funds.

18

Healthier homes: the role of health authorities

Martin Bardsley

18.1 Introduction

This chapter will look at how cold housing affects general health, and in particular the role of health authorities in tackling this problem. The general relationship between housing and health has been acknowledged for some time and some of the earliest pioneers of public health recognised that improved quality of housing is one means to improve the health of the population. Yet the responsibility for designing, building and maintaining housing stock lie with a different set of organisations from those that are directly responsible for dealing with adverse health effects. For this reason health services are interested in playing their part towards the development of better quality housing as one means to improve the health of the population - though it is not always obvious what role the NHS can play. Such approaches are consistent with the current government's emphasis on inter-sectoral work and the infamous 'joined-up thinking' as a way of tackling problems in one sector through action in another [Department of Health, 1997a].

" Over recent years it is the health and well being of people living in the most run-down communities which have suffered most. Poverty, low wages and occupational stress, unemployment, poor housing, poor education, limited access to transport and shops, crime and disorder and a lack of recreational facilities all have had an impact on health."
[Department of Health, 1999]

As well as the strategic considerations, there are also a number of practical operational situations where housing and health services come into contact. This could be in the community care team looking to discharge an older person from hospital into accommodation that will not further damage their health. Or, it could be the local GP seeking re-housing or refurbishment for a patient whose housing plays a major contributory factor in their ill health. It is for these reasons that health and housing sectors interact in different ways. The specific issues relating to heating are just one part of the wider questions about housing and health. This chapter will look at some of the ways this is happening and some of the problems that have to be overcome.

18.2 Cold damp housing and health

As this book indicates, a growing research base shows how cold and damp housing can a have a negative effect on health. Several specific aspects of housing have an impact on health status, including cold and dampness, but other factors to consider are overcrowding, indoor air quality, design, noise and infestation [Hunt 1997; Pollard, 1997; Ambrose et al..., 1996, Burridge and Ormandy, 1993; Standing Conferences on Public Health, 1994].

Problems with damp usually arise from a combination of poor quality construction, cheap materials and inappropriate renovations, with a lack of insulation and adequate ventilation, combined with expensive and inappropriate heating systems [Hunt, 1997]. The effects of cold are also closely related to those of damp, as lower temperatures in housing encourage condensation, mould growth and the proliferation of fungal spores. Previous chapters have described studies which have linked cold and damp to a range of health problems (see also Box 18.1 and, for example, Platt et al.. [1989] and Hyndeman [1990]).

These physical health problems are exacerbated by the problems of those on low incomes. Around 8 million households experience 'fuel poverty' and cannot heat their homes adequately [Rayner, 1994; Boardman, 1991]. The effects of cold are felt to be most marked amongst older people, partly because of the lifestyle they lead which tends to be more sedentary, but also because their age may make them more susceptible to ill health [Boardman, 1991]. In this group as well, there may be the interaction of low income, poorly maintained homes.

Accidents, falls, fires
Exacerbates respiratory problems including asthma and infections
Allergic reactions from fungal spores
Cardiovascular disease including heart attacks and stroke.
Also reduces peoples ability to cope with pre-existing conditions
Mental health problems including stress
Lower levels of general well-being

Box 18.1. Range of health effects of cold and damp housing

18.3 Use of NHS resources

In addition to the effects on health, there are also demonstrable effects on the use of health services raising the prospect that investment in housing services may be one means of reducing, or at least changing, the demands on health services. In fact poor quality or no housing, affects a number of other sectors of government and the local economy. There is obviously an interest in the extent to which investment (or the lack of it) in one sector may have financial implications in others. Some studies have tried to quantify these implications and one suggested that in 1991, condensation cost the NHS around £800m a year [Burrows and Walentowicz, 1992]. Another estimate suggests that the costs to the NHS of treating ill health resulting from sub-standard housing is estimated at £2.4 billion a year [National Housing Federation, 1997].

Other studies have looked in detail at how housing conditions affect both health status and the use of NHS resources. Table 1 gives an example of how indicators of poor health are worse amongst people in the worst housing. Carr-Hill and colleagues [1993] looked at 203 adults on one poor estate and compared on a series of health indicators to a matched sample from the General Household Survey. The excess morbidity appears in most age/sex groups, but is especially marked in older people. This study also found that the costs of health services were approximately 50% higher in the sample of people from this estate, i.e. higher than a sample matched for income. It suggested that if these results are extrapolated to the 10% of houses with damp [Department of the Environment, 1993], the costs to the NHS would be around £600m in 1994.

Table 18.1. Comparing observed to expected health status for those in poor quality housing

	Ratio of observed/expected cases*		
Women	*Age <44*	*Ages 56-64*	*Ages 65+*
% in 'poor health'	4.50	1.87	3.21
% with long standing disabilities	1.48	1.39	1.33
% with limiting long term illness	1.12	1.31	1.28
Men	*Age <44*	*Ages 56-64*	*Ages 65+*
% in 'poor health'	1.00	2.18	1.76
% with long standing disabilities	0.96	1.02	1.13
% with limiting long term illness	0.88	1.51	1.24

* Expected values based on matched populations in the General Household Survey   *Source: Carr-Hill, Coyle and Ivens [1993]*

More recently, as Peter Ambrose describes in more detail (Chapter 17), Barrow and Bachan [1997] conducted a similar study in 1996 of 107 households on two estates in East London. In terms of health service costs, the study concluded that residents on these estates cost approximately £515 per year compared to £72 per year in an improved estate in Paddington. The biggest contribution to these costs was in-patient stays.

These figures are inevitably imprecise. However, they point to substantial annual costs to the NHS that can be associated with poor quality housing. These do not include any excess costs associated with institutionalisation. It is important to consider therefore how investment in housing might yield cost savings, recognising that the pay-off from investment is spread over many years. It is also important to remember that poor housing has its costs in other sectors of public expenditure, for example education and crime as Barrow and Bachan describe. If the annual revenue costs are considered in terms of equivalent capital values, then these values increase by at least ten-fold.

18.4 Understanding the evidence

One of the biggest problems for health and housing sectors is in understanding how the evidence on quality of housing should be affecting the decisions that are made. In the NHS, as elsewhere, decisions about how to use finite resources should be based on good evidence of effectiveness. In practical terms, local health authorities need to be able to say what health benefits will accrue from investment in housing – something that is not always easy to do in a local situation.

Perhaps the biggest problem is dealing with the complexity of the relationship between cause and effects and to disentangle individual factors in the housing environment [Carr-Hill, 1997]. Firstly, poor quality housing itself has many effects and there are no simple health measures that act as a unique barometer of change. Box 18.1 shows the range of health effects linked with cold damp housing. Moreover, as later sections discuss, many of the routine health indicators that we have are not very good at identifying these different health problems.

Secondly, the health effects are not specific to cold and damp housing but are associated with a wide range of other factors. For example, the association between poor quality housing and poverty is inevitable and poverty itself is associated with poorer health. Among London boroughs there is a strong relationship between deprivation (using the Jarman score).and a common indicator of community health status (standardised death rates below age 75). At the extremes, the chances of dying before the age of 75 are 50% higher in some boroughs than in others. This relationship exists for most health indicators and across other measures of deprivation. In order to separate out the effects of housing, approaches are needed that adopt either standardisation for confounding factors or statistical methods for dealing with multiple variables.

Finally, much of the research evidence is based on small-scale studies in particular areas. This inevitably begs the question about the extent to which these results may be generalized. Moreover, cross-sectional studies are able to demonstrate associations between housing and ill health, but do not necessarily say what is effective in improving health. In particular, there is a shortage of studies looking at how marginal change in housing leads to marginal changes in health status. One of the critical questions for directing action is not whether there is a link between housing and health, but rather what are the health benefits of changes in the condition of people's housing. This is methodologically a more complex problem and requires a significant investment in the right forms of research.

18.5 Using local health information

One of the first steps for local health agencies is to consider how poor quality housing may play a part in local profile of health. There are a range of health indicators that are potentially of value, for example in the Public Health Common Data Set [Department of Health, 1997b], but their interpretation is complex. These indicators are primarily of use in developing locality health profiles rather than the specific identification of particular housing problems. Local health indicators can also help in disseminating messages about health and raising awareness of issues. Table 18.2 summarizes some of the potential health information that might be used.

The effects of cold are seen most directly in cases of hypothermia. Reference has been made in Part One to the estimated extra 8,000 deaths for each degree Celsius the temperature falls below the average in cold weather. Deaths from hypothermia are more common in the elderly, with an average age at death of 77 in men and 81 in women. Overall, deaths coded as hypothermia form only a small percentage of deaths and rates have been falling [Chantler and Kelly, 1999]. The very low numbers mean that it is not really useful to consider local rates on a routine basis.

Cold will also have effects on respiratory function and the risk of heart disease [Leather et al.., 1994; Ineichen, 1993] which is assessed through measure such as the excess number of winter deaths (Chapters 3 and 4). The numbers of excess winter deaths have long been recognised as a health status indicator, typically measured as an index representing the number of deaths between January and March relative to the all year average. At present this index stands at around 1.11, although it is

Table 18.2. Potential sources of NHS information that may be used to look at housing conditions and health

Information	Comments
Mortality	Widely used and available for mapping to small areas, but the smaller the areas the rarer the events. Major problems understanding how causes in the housing environment contribute to observed mortality rates, given the range of health determinants and the time lag that may exist between an intervention and the eventual outcome.
Hospital inpatient data	Widely available by diagnostic category and for small areas. Has to be treated with some caution as hospital admission rates may say more about the provision of care locally rather than the underlying prevalence of disease. Hospital admissions also represent the 'severe' end of health spectrum. The problems of linking multiple causes to multiple effects present problems similar to that seen for mortality.
Infectious disease notification	Notification is made by GPs for some selected conditions, some of which are related to housing conditions eg tuberculosis. Problems are that not all infectious diseases are covered and that reporting can be selective in some areas.
GP activity	Not available in detail on a routine basis, but with co-operation from local practices it is possible to use activity data, e.g. visits to GPs, and diagnostic data, e.g. visits for asthma, as part of local health status measures. In some areas there may be reasonably complete local registers for certain chronic disease such as diabetes or asthma.
Survey information	National surveys, such as the Health Survey for England, or General Household Survey, do not have sufficient sample sizes for local monitoring. In many cases health authorities may undertake local ad-hoc surveys. Some general health status tools asking people about their perceived health have potential as tools for looking at housing related changes in health.
Accident data	Apart from mortality and hospital admissions, there are two main sources of accident data – road traffic accidents collected by the police [London Accident Analysis Unit, 1997], and a national survey [Department of Trade and Industry, 1995]. Though there are no routine data sets that use information from A&E departments, these can be used, with co-operation, to look at specific issues for those attending A&E.
Activity in community health services	The NHS collects a number of statistics describing activity by community health services, usually grouped according to trusts. These are primarily about contacts with professionals and as such the extent to which they reflect local health status is limited. Specific research projects, in collaboration, may wish to use the experience of groups such as health visitors or district nursing staff as part of data collection exercises.

sensitive to annual variations according to the severity of the winter [Curwen, 1997]. There is a specific published indicator [Department of Health, 1996] relating to those aged 75 and over. The highest values in the country are over 1.3, the lowest around 1.0. International comparisons suggest that the values in England and Wales are high compared to most other countries (Chapters 1 and 2). The interpretation of differences in excess winter deaths is complex [Donaldson and Keatinge, 1997] and can be affected by influenza epidemics [Curwen, 1997] and external temperatures.

Table 18.3 summarizes the values for health authorities in London and shows the average over three years. It should be noted that the confidence intervals around any one value tend to be large and many of the differences are not significant and there is no very clear pattern between areas in London. One might expect that such an indicator would be consistently higher in the inner city, but this is not necessarily the case.

Table 18.3. Excess winter deaths ages 75+, London health authorities 1993-95
Source: Modified from Department of Health, 1996

Health Authority	Ratio Winter:Annual	95% CIs Lower	Upper	Approximate number
Barnet	1.13	1.09	1.18	365
Hillingdon	1.23	1.17	1.29	409
Brent and Harrow	1.18	1.14	1.23	591
Ealing, Hammersmith and Hounslow	1.23	1.19	1.26	972
Kensington and Chelsea and Westminster	1.22	1.17	1.27	503
Barking and Havering	1.21	1.17	1.26	712
Redbridge and Waltham Forest	1.25	1.2	1.29	977
East London and The City	1.24	1.20	1.29	862
Enfield and Haringey	1.24	1.20	1.28	854
Camden and Islington	1.21	1.16	1.26	515
Bromley	1.16	1.11	1.21	424
Bexley and Greenwich	1.22	1.18	1.27	810
Lambeth, Southwark and Lewisham	1.16	1.13	1.20	816
Croydon	1.23	1.18	1.28	539
Merton, Sutton and Wandsworth	1.20	1.17	1.24	1,012
Kingston and Richmond	1.17	1.12	1.22	488

The conditions that lead to excess winter deaths are also likely to be part of the increased pattern of winter admissions to hospital. As with excess deaths, there are many factors that will affect admissions including the supply and use of local hospital beds. Hospital admission data are widely available and can potentially be mapped to specific areas such as wards or enumeration districts (Chapter 10). Figure 18.1 presents a summary based on admissions from North East London suggesting that for respiratory and ischaemic heart disease, admission rates in winter months (November to February) are significantly higher than in the summer months. The effects are most marked in young

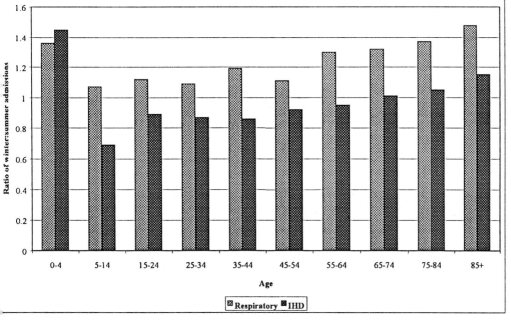

Fig. 18.1. Ratio of winter/summer admissions to hospital for respiratory disease and Ischaemic Heart Disease. *Source* Bardsley *et al. 1998*

children (under age 5) and older people, where rates in winter are 40-50% higher than in summer.

However, cardiovascular disease is a major cause of both ill health and mortality – it accounts for half a million deaths per year in the UK and encompasses two broad groups of diseases Coronary heart disease (CHD) and cerebrovascular disease (CVD or stroke). Cardiovascular disease has been a target in national health strategies [Department of Health, 1999]. Though the incidence of CHD is not known precisely, estimates suggest that around 80-90 cases per 10,000 population are diagnosed each year [Langham *et al.*, 1994]. Cerebrovascular disease (stroke) is also a relatively common health problem, for every 100,000 people there are an estimated 240 acute strokes per year, and around 600 survivors alive in the community [Wade, 1994]. There is a great deal of research into the key factors increasing the risk of heart disease such as diet, smoking, physical activity etc. The effects of housing must be considered as secondary to these. Poor quality housing may be an additional complicating factor for the many people with pre-existing cardiovascular disease.

This group of diseases forms a significant proportion of all hospital use - London records more than 100,000 admissions per year. At ward level there appears to be a significant association between CHD admission rates and the proportion of houses lacking amenities, ethnicity and unemployment, but not with the levels of overcrowding [Bardsley *et al..,* 1998]. Respiratory diseases are one of the most common forms of ill health. Asthma is estimated to affect around 19% of schoolchildren by the age of seven years, and one in four to one in five sixteen years olds report respiratory symptoms related to asthma [Prescott-Clarke and Primatesta, 1997]. Around 11% of adults have asthma as diagnosed by a doctor. It is thought that there has been a rise in diagnosed asthma over the past few years, although debate rages about the extent of the change and the reasons for it. Nevertheless, there does appear to have been an increase in the use of health services in relation to asthma. In London there are more than 80,000 admissions to hospital every year for respiratory disease (coded as ICD9: 460-519), equivalent to about 12 admissions for every 1,000 residents. Rates are highest amongst men and older age groups (55+). Respiratory diseases are a major cause of death, especially in the elderly. Approximately 12% of all deaths are from respiratory disease, mainly pneumonia, bronchitis and emphysema.

There are a number of factors in the home environment that can exacerbate respiratory disease including cold, damp, indoor air quality and the effects of house dust mites (Chapters 3 and 5). Despite the significance of chronic respiratory conditions such as asthma or bronchitis, there are no routine NHS information systems that enable us to look at the differing prevalence [Hyndeman *et al.* 1994]. Although there are some local analyses, e.g. those based on local health surveys or around individual General Practices or hospitals, our main sources of information across London have to be based on the most severe manifestations of illness, i.e. when people are admitted to hospital or die.

The effects on mental health at a local level are even harder to unravel. Research studies have shown that mental health problems are greater among those living alone [Shapiro *et al.*, 1985;Goldberg 1990], and that socio-economic factors in an area (some of which relate to housing) are associated with levels of mental health admissions to hospital [Glover, 1996; Kammerling and O'Connor, 1993]. Though information on hospital admissions is widely available, it suffers from a number of problems especially when the relationship between hospital use and local needs are affected by patterns of local provision and practice.

For infectious disease monitoring there are systems for selected conditions. However, the notifications systems can be dependent on the interest and enthusiasm of individual doctors and housing conditions may be only one contributory factor to the overall incidence.

18.6 Health service organisation and policy

There are different branches of the health service that can relate to housing. Over the past twenty years the NHS has experienced a succession of organisational changes that have not helped those outside the service know whom they need to be speaking to. The providers of face to face clinical services are in the health trusts, which may be local hospitals or community services, or by primary care teams based around a General Practitioner. The planning and funding of services is based on money allocated to health authorities which will then be distributed to providers by a variety of mechanisms, in some cases through the newly emerging Primary Care Groups (PCGs).

The NHS White Paper *The New NHS: Modern* and *Dependable* [Department of Health, 1997a] required the development of local Health Improvement Programmes (HImPs) which aim to look at the broader determinants of health and charge health and local authorities with specific duties in relation to collaboration. The implications of the White Paper are that health authorities become more focused on strategic health improvement. Primary Care Groups will consist of groupings of GP practices for populations of around 100,000, which will take over responsibility from Health Authorities for commissioning acute and community services for their population. The proposals represent a major shift in the way services within the health service are planned and co-ordinated. There are several potential stages of development for PCGs, ranging from acting as advisory bodies to the health authority to integration with community service providers. The development of PCGs as the lead in local commissioning may have advantages and disadvantages in this respect. The greater role for family doctors and other primary care professionals may help in forging the operational links to local housing departments. The danger is that the geography of PCGs, and the catchment populations they deal with, may make it harder for housing agencies to know whom they should relate to. Similarly, the management component required for developing local plans, including some of the public health work on needs assessment, is largely unknown. It also important to recognise that it will take time for the process of change in management structures and the development of the PCGs to settle down and for the benefits of these new arrangements to emerge.

The latest national health strategy *Saving Lives: Our Healthier Nation* [Department of Health, 1999] places an important emphasis on tackling health inequalities and the wider causes of ill health. Housing related problems are included and there are some specific government initiatives that are felt to be capable of bringing health benefits (see Box 18.2). The strategy sets targets in four areas, cardiovascular disease, cancer, accidents and mental health – three of these have a significant cold housing dimension. Although housing is considered to be one of a range of other determinants of poor health., it is not explicitly mentioned as one of the target areas. The idea that a handful of national targets be supplemented by local specific ones offers the potential for thinking about how housing related targets can be considered in local plans. The government is also placing an emphasis on Health Impact Assessment (HIA) which has been defined as *'the estimation of the effects of a specified action on the health of a defined population'* [Scott-Samuel, Birley and Ardern, 1998]. This technique goes beyond the description of health states to a prospective estimation of how these will change as result of a particular policy or programme and as such is well suited to use in the context of housing initiatives. However it must be admitted though the term Health Impact Assessment is commonly used, there are still relatively few such analyses that have been carried out in practice and there is as yet no established methodology.

Objectives of the national strategy
- To improve the health of the population as a whole by increasing the length of people's lives and the number of years people spend free from illness
- To improve the worst off in society and to narrow the health gap

Specific housing related initiatives:
- Campaigns to address fuel poverty and improve energy efficiency such as the *Keep Warm, Keep Well* campaign for older people
- Reviewing the housing fitness standard to include health and safety including poor energy efficiency
- Housing investment including the National Housing Investment Programme and Capital Receipts Initiative; Approved Development Programme for the Housing Corporation; Private Sector Disabled Facilities grants; funds for Home improvement agencies
- New Home Energy Efficiency scheme
- Initiatives to reduce homelessness with a target to reduce people sleeping rough by two thirds in 2002

Box 18.2. Objectives in saving lives   *Source: Our Healthier Nation [Department of Health, 1999]*

The significance of housing as one of the factors leading to health inequalities was explored in the 1998 Acheson Report which included a specific recommendation for policies *'to improve insulation and heating systems in new and existing buildings in order to further reduce the prevalence of fuel poverty'* [Acheson, 1998]

Health Action Zones (HAZ) represent a pilot scheme for forging new relationships between agencies with a role in health improvement and reducing inequality. The first wave in April 1998 included 11 schemes in some of the most deprived areas of England. These have been supplemented by a further 15 second round bids in 1999 and a total of £290 million is being made available over three years. The different HAZ schemes are using different approaches and focussing on local priorities. The significance of the Health Action Zones is the idea that in these areas with particular health problems, there will be new flexibilites that will encourage better co-ordination between health and local authorities.

One potential barrier to 'joined up' working across sectors is that there is no simple mapping between housing agencies and health agencies. In some areas where a single health and local authority are coterminous, there are distinct advantages when compared to the health authority that needs to relate to three different local authorities. In most boroughs there will more than one Primary Care Group. Similarly, Housing Associations and the voluntary sector may work to different geographic areas altogether. These differences in organisational structure and geography are not insurmountable but they do not help in forming partnerships.

18.7 Elements of work by Local Health Authorities

The problems posed by poor quality housing are dealt with in the short term by a combination of agencies - environmental health officers, housing officers, social landlords, social services departments, private landlords and individuals, the emergency services, GPs, community health staff, etc. For health authorities to play a role there are needs to be a degree of shared understanding and co-operation between agencies.

One of the problems in fostering inter-agency co-operation is that responsibility for housing-related issues tends to be fragmented within the NHS and even within individual health authorities. For example, a health authority will typically have different people responsible for medical priority re-housing, work on joint finance, community care, primary care development, health needs assessment, research and/or high level strategic links to regeneration initiatives etc. Moreover, local trusts will have different people again linking to local housing authorities and social services and in primary care, whilst GPs will have a different set of relationships to the local council or community groups. It may be that this fragmentation is partly the result of a failure to recognise the strategic importance of the links between health and housing.

There are a wide variety of ways that the work of local health authorities can contribute towards healthier housing. Table 18.4 (over page) gives an example of the breadth of issues that might be covered and by an individual health authority. The following sections discuss some of these in more detail.

Table 18.4. Auditing health authority activity in relation to housing

Themes	Specific pointers	Why?
Strategic planning	• Is health considered in local housing plans? • Is housing part of the Health Improvement Programme/health strategy? • What contact do you have with LA housing leads? • What contacts with other housing organisations? • Are there any joint planning posts? • Do you have any partnerships relating to housing?	• Major health gains are possible through improved quantity and quality of housing • Health service planning, especially for vulnerable groups, depends on availability of appropriate housing • Changes in the patterns of housing can have knock-on effects on needs for and use of health services
Specific housing-related projects	• Funding of house adaptations/improvements? • Funding of work to improve access to health services for homeless people? • Programmes for developing NHS staff roles in recognising, recording and dealing with housing-related problems? • Co-ordinated planning of primary health care facilities with housing regeneration/new building? • Funding of programmes for increasing awareness amongst NHS staff of housing problems and their links to health?	• Reduce inappropriate demand for community care services • Enable prompt discharge from acute settings for some client groups • Reduce demands for emergency or acute services for homeless groups through better promotion, prevention, diagnosis and early treatment of illness • Promote better health through wider environmental changes in specific areas and improving general quality of life for residents • Ensure most appropriate treatments for health problems relating to housing
Medical priorities	• Are medical priorities for re-housing dealt with appropriately? • Are there explicit criteria and protocols for decisions? • How are housing providers involved in decisions? • How are GPs involved in decisions? • What interventions are available, other than re-housing? • Are protocols agreed with local authorities?	• Poor quality housing can have serious effects on people's health. With limited supply it is important that re-housing is offered to the most urgent cases • GPs can become disaffected with a system that does not appear to be effective. Housing agencies can become disaffected with a system that is unrealistic given the stock available, inconsistent or open to abuse.
Community care	• How well do community care assessments recognise the housing dimension? • How effective are discharge planning procedures? • How well do health and community care priorities link to housing plans?	• Availability of appropriate housing is essential to community care • Appropriate discharge planning can reduce unnecessary use of the acute sector
Health promotion	• Are you ensuring work in relation to accident prevention, energy efficiency, indoor air quality and problems of poor quality housing? • Are there programmes to improve access for the homeless or those in temporary accommodation?	• Effective prevention of health problems related to poor quality housing can improve the health of individuals and offer more appropriate use of NHS resources
Health information and needs assessment	• Do you know the health services priorities with regard to housing in your area? • Have you done any work on health needs in relation to housing? • Do housing-related issues feature in the Director of Public Health's Annual Report? • Do you provide health information to local housing agencies?	• Joint working and planning needs to consolidate on joint information • Housing agencies need to be aware of health priorities and ways to use health service information in planning • Understanding the health issues in relation to housing is an important pre-requisite for joint working

18.8 Assessing health and housing needs

The health authority has an important role to play in developing health information and assessment that can be used in the housing sector. This might take the form of including housing specifically in Public Health reports or health strategies. For example, Bexley and Greenwich Health Authority's Annual Public Health Report 1996 talks of the links between local health indicators (excess winter deaths) and housing conditions, in particular cold and damp. The report identifies a number of ways that the local health services can improve their work with the housing department in tackling these problems (see Box 18.3).

- Housing departments should target initiatives to improve energy efficiency and install central heating for the most vulnerable groups
- Housing departments should work with the NHS to identify vulnerable groups and publicise ways of improving energy efficiency and heating costs for them
- Joint local authority/health authority/health promotion initiatives should publicise adequate heating levels for health, ensuring that vulnerable groups are consulted
- Grants for home improvements to private households should be targeted on those homes with lowest energy efficiency ratings
- Scope for further work to monitor and prevent home accidents

---

Box 18.3. Example of the recommendations relating to housing and health in the Report of Directors of Public Health

There may also be a sharing of information about health status or service utilisation with local housing agencies. Public health departments can help in providing information and interpretation of local health data as well as ways of using research literature to best effect [Aspinall, 1994; Marchant, Moore and Ditchburn, 1996]. This may include work with housing departments to identify vulnerable groups, or promoting energy efficiency to reduce heating costs. Croydon Think-Tanks [1995] were a multi-agency exercise looking at broader determinants of health, and included looking at housing. The work helped in identifying priorities for housing improvement work and a new project looking at home safety and security for elderly residents.

One example is work in the London Borough of Hounslow where on one estate, a public health officer was employed jointly by the health and local authorities. The aims of the post were to prioritise needs, and establish information and feedback systems, and to develop an action plan and public health targets. The role of the Health Visitor can be an important way to identify health problems related to housing. A number of health authorities are involved in developing health visitor roles to specifically link to housing problems.

It must be accepted that we are still some way from a common language in which health benefits can be linked to housing. The previous section pointed to the problems of linking health information to housing and the uncertainties over what information is most useful. There is important developmental work to be done in establishing just such a common framework for discussing housing and health.

ASSESSING MEDICAL PRIORITIES:

People can apply to the local authority for priority status on medical grounds and such applications are usually supported by their GP. However, the ability to rehouse people on medical grounds is increasingly limited as demands have grown and the available council stock has decreased. A relatively large proportion of those who claim medical priority fail to be re-housed or have to endure long waits. The evidence about the health outcomes is mixed, with better improvements linked to mental health problems [Elton and Packer, 1986] than physical illness [Golding, 1987]. Local authorities have varying mechanisms for designating medical priority, sometimes using their own advisors, sometimes seeking the advice of Public Health professionals in the health authority. There are some examples of where collaboration between health and local authorities on priorities is well developed. However, on other areas there are questions about how well, in general, the work on medical priority re-housing links to either a wider public health agenda or community care. [Smith *et al.*, 1993]. One study [Easterlow and Smith,1997] has reported that the rationing of medical and mobility priority was causing frustration among health advisers. Many areas are moving towards more systematic assessments of medical priorities and placing a greater emphasis on communication between health advisers (including GPs) and local housing services (see Box 18.4).

- Develop communications and joint working with housing providers
- Recognise range of interventions (other than re-housing) that may be appropriate
- Encourage more standardised systems of assessment
- Promoting ways of incorporating medical assessments fairly in housing waiting lists/priority systems and better monitoring of medical priority allocations
- Accountability and consistency into how 'health needs' are built into access and allocation procedures.

Box 18.4. Approaches to improving medical priority systems

HEALTH PROMOTION

This is based in either health authorities or local trusts and can include a recognition of housing and heating issues in initiatives. Such work may involve campaigns with the public and professional staff to raise awareness of the problems posed by poor quality housing with NHS staff. In practice, work around heating and health may overlap with a range of other issues specific health problems, such as accidents, or relate to particular client groups, such as elders or the homeless [Drennan and Tearn, 1996]. It is important that such work is linked in with local housing agencies, advice centres and GPs and health centres.

JOINT INVESTMENTS IN HOME ADAPTATIONS OR IMPROVEMENT

A genuine sharing of budgets across organisations is often one of the most difficult aspects of partnership working to achieve, and there are underlying differences that result from distinct budgets and lines of accountability. Thus, planning and finance cycles in housing and health services do not necessarily match. Though this may not be an insurmountable difficulty it does make life more difficult. There is also an inevitable defensiveness within an organisation when it comes to issues such as finance and control. Why should health services pay for housing improvements when these are the responsibility of housing providers? Molyneux and Palmer [1997] suggested that there are 'cost drivers', which tend to increase costs to the individual and society, that result from existing ways of organising and financing housing and social care. They argue that existing health and community care plans need to focus as much on joint commissioning and investment as on planning.

For health services, whilst the logic of investing in housing to prevent ill health may be sound, the practicalities of diverting significant funds from budgets that are always under strain make large scale direct investment unlikely. Though access to NHS resources may be attractive for some in the housing sector it does raise important questions about roles and responsibilities of different agencies of government. Yet, as demonstrated in Chapter 14, for example, there are instances of where health agencies have invested resources in housing. In one area, money provided by central government to reduce winter pressure on hospitals has been spent on housing improvements to reduce the hospital admissions. In other areas, health authorities have provided some funding for installation of central heating systems for people with respiratory problems. Some health authorities have worked on schemes for funding minor repairs/modifications. There are 20 different funding mechanisms for aids and adaptations involving GPs, health authorities, social services, housing associations, district councils and The Housing Corporation. Kirklees Council have developed a service that brings together different services to deliver adaptations referred by social services [Association of Metropolitan Authorities, 1997]. For health services, the links between housing adaptations and discharge from hospital are a key factor in community care and, in particular, in discharge planning for hospitals.

REGENERATION INITIATIVES

The association between poor housing, poor health and a whole battery of other social indicators mean that health improvements are most likely to come from wide ranging inter-sectoral developments such as those seen in the various urban regeneration schemes.

"There is now a greater awareness of the importance of a more rounded holistic approach to housing regeneration. Physical refurbishment is seen as only one aspect of improved housing quality - to be fully cost effective housing investment strategy must be allied to additional investment in social and economic regeneration" [Ambrose et al.., 1996]

Although health should be an important consideration for decisions over local planning and development issues, the extent of explicit consideration of health factors tends, in practice, to be limited. Thus, for example, Health Impact Assessment, which should be ideally suited to appraisal of these investments, is still relatively untested and lags behind Environmental Impact Assessment [British Medical Association, 1998]. The relative underdevelopment of health assessment in these settings stems partly from the fact that, historically, communication between different organisations or sectors has been limited. The second reason is the problem of establishing a credible, practicable and valid way of describing health that can be used in other sectors.

To date, the extent to which health-related criteria are used in developing and assessing Single Regeneration Budgets (SRB) bids has been limited and before 1997 health authorities have not tended to be major partners in bids. However the bidding guidance for the 1998 SRB round does include mention of public health and suggests bids that are consistent with national policies and programmes, amongst which is public health [Department of the Environment, Transport and the Regions, 1998]. The roles that health authorities play in such bids is increasing however, and it is now commonplace for health themes to be included, and sometimes to lead, major regeneration bids. A similar engagement with health authorities is also need on a range of other government initiatives, for example, the New Deal for Communities approach to neighbourhood renewal [Social Exclusion Unit, 1998]

EVALUATION

Information on the health benefits of regeneration strategies is fairly limited. This makes it harder for inclusion of a health dimension in such bids. It is important that health criteria are included in the overall assessment of the success or otherwise of such schemes. The types of health indicators that might be used range from the relatively sophisticated to generic quality of life or health status measures that might be completed by tenants on a before or after basis. In addition, information about health service use might be brought to bear (see example from Holly Street, Box 18.5)

- *Evaluating housing improvement in Stepney:* Part of the local investment in housing improvement has included an evaluation of health status in a before and after design. This is also linked into a wider assessment of the costs of poor housing in terms of education, crime and other services [Ambrose *et al.*, 1996].
- *Holly Street, Hackney:* A major regeneration project of nearly £100 million, which includes over 1,000 new homes, community facilities and a primary health care centre. There have been two surveys of residents undertaken so far and another is planned. Results to date show improvements in residents' perceptions and experience of crime and community safety, and that the new homes are considered healthier places to live. Surveys also suggested that residents in new properties were using health services less [East London and The City Health Authority, 1998].

Box **18.5.**: Evaluating health impacts of regeneration

18.8 Making the strategic links

In the longer term, there are a variety of mechanisms for improving the stock of existing houses such as House Renovation Grants, capital expenditure on energy conservation, or for new buildings there are a number of design standards that are either in use or being proposed. For health services it is important to realise that investing in good quality housing now can reap longer term benefits [Ambrose *et al.*, 1996, Molyneux and Palmer, 1997].

For the strategic development of better quality housing there is a role for health authorities with other agencies in ensuring that health messages are put into major regeneration programmes and priorities for local housing investment. Yet we have to recognise that such collaboration will not necessarily happen spontaneously. There are a number of approaches that have been used in improving the links between different organisations, for example:

- programmes for sharing knowledge about different sectors with a willingness to improve communication and understanding about the structure, problems and priorities of different organisations;
- shared staff between health and local authorities can help overcome some of the problems associated with understanding different organisations;
- strategic level joint meetings which encompass both sectors and can give a sense of common purpose to joint initiatives;
- developing a shared R&D agenda, with particular emphasis on the health benefits that accrue from housing investment;
- making the explicit links to health outcomes as part of housing investment programmes or regeneration bids;
- sharing information about health and housing conditions;
- linkages between Health Improvement Programmes and the local housing strategies, and vice versa.

It is important to be clear about the advantages to the health service of a greater involvement in housing issues. There are differences between, say, the long term health gains that result from regeneration work, contrasted with the benefit of operational links between discharge planners and those responsible for housing adaptation. Part of this relates to improving understanding of housing amongst the health sector - and vice versa.

The recognition of the common strategic links between health and housing organisations is an important precursor to effective joint work. The development of Health Improvement Programmes, following the NHS White Paper [Department of Health, 1997a] offers an opportunity to build these links into strategic planning within the NHS. However, an exhortation to health and housing authorities that they 'can do better' has to recognise that both groups need to be clear regarding the benefits of further investment of often limited managerial resources in joint work. It also needs to recognise the good work that may be going on at present.

19

The unavoidable imperative: cutting the cost of cold

Peter Smith

The link between poverty and ill health has long been obvious, but it was not until Sir Douglas Black produced his report on the subject in 1980 that it became official. The preceding chapters underline the fact that poverty associated with poor housing is a major contributory factor to illness and premature death. This is now acknowledged in a Government document *Fuel Poverty: The New HEES* [DETR 1999] which defines fuel poverty as households which have to spend over 10% of income on fuel "to maintain a satisfactory heating regime" plus adequate lighting and the running of typical appliances. According to the 1996 English House Condition Survey, [DoE 1996] at least 4.3 million households come into this category. The DETR document goes on to say:

"The principal effects of fuel poverty are health related, with children, the old, the sick and the disabled most at risk. Cold homes are thought to exacerbate existing illnesses such as asthma and reduced resistance to infections." [DETR 1999]

The various contributors to the book remove any doubts there may be about the linkage. Increasingly damp as well as cold is emerging as a major health hazard. Damp generates mould and mould spores can trigger allergies and asthma attacks. Some moulds are toxic, as in the genus *Penicillium* which can damage lung cells. It was confidence in the connection between damp homes and asthma that justified the Cornwall and Isles of Scilly Health Authority in directing £300,000 via District Councils to thermally improving homes of young asthma patients (Chapter 12). This was undertaken as much as an investment opportunity as a remediation intervention. The outcome was that the savings to the NHS exceeded the annual equivalent cost of the house improvements.

The report on this enterprise, sponsored by the EAGA Charitable Trust, states: *"This study provides the first evaluation of health outcomes following housing improvements"*. [EAGA 1999] It will surely be the first of many since it provides hard evidence of cost effectiveness. At the opposite end of the country, almost a quarter of all homes suffer from damp in Scotland. (National Housing Agency for Scotland).

How does the figure of 4.3 million fuel poor in England relate to the housing stock condition? The International Energy Agency which considers energy efficiency worldwide described UK housing as 'poorly insulated' with 'considerable scope for improvement'. At the same time, government improvement programmes were 'unconvincing' with 'funding low in proportion to the magnitude of the task'. What is the magnitude of the task?

The government criterion for energy efficiency is the Standard Assessment Procedure (SAP) which comprises a calculation of the heat loss resulting from the form of the building, the thermal properties of its fabric and the level of ventilation. This is equated with the cost of making good the heat loss through the heating system and the cost of fuel. There is compensation for solar gains. New homes complying with Building Regulations according to the SAP method have to achieve SAP 75 plus. The recommended minimum for reasonable energy efficiency is SAP 60. The English House Condition Survey found that 86% of dwellings were below SAP 60 with 8% below SAP 20 and 30.35% under SAP 30.

The current average for England overall is SAP 44. This is gradually improving as the ratio of new homes to existing increases. However, in the private rented sector in England the average is SAP 22, with 21% of this sector being below SAP 10. Within the <10 category the bottom end is as low as SAP *minus* 25.

In the face of this enormous burden of dereliction, the government allocation of £300 million between 2000 and 2002 will barely make a dent in the problem. A KPMG report of 1998 [KPMG 1998] estimated that it would need about £5000 per dwelling to achieve a standard of around SAP 65 through insulation including double glazing and overcladding. This is broadly consistent with the 1991 English House Condition Survey estimate of £4860 per dwelling to a level of SAP 60 [DoE, 1996].

In Penzance a recently completed upgrading programme by the Penwith Housing Association bears out these figures. The Association took over the local Council's housing stock, a third of which was solid wall construction from the 1930s. The worst cases were as low as SAP 10. The transfer involved raising private funding to bring the stock up to a state of good repair and thermal efficiency over five years. This included external insulation to all solid wall houses (see Fig. 19.1). The full improvement package of overcladding, roof insulation and double glazing raised the standard to SAP 76, equivalent to a National Home Energy Rating (NHER) of 8.2. The cost of the external wall insulation for mid terrace dwellings was £4290. End of terrace costs were double this figure. [PHA 1999] As expected, the tenants have taken much of the benefit in improved living conditions, with reports of considerable gains in quality of life.

Fig 19.1. If there is to be a serious attempt to tackle poor housing starting at the <SAP 20 end of the spectrum, it is clear that costs have to be considered which are several orders of magnitude greater than the government's current commitments.

A funding model

The KPMG [1998] model assumed an upgrading cost per dwelling, as mentioned, of £5000. It also incorporates the mechanism of the Energy Service Company (ESCO) as the means of delivering both energy and the thermal upgrading. The Government definition of an ESCO is: "a company providing a complete energy service, i.e. combining energy supply with the provision of measures concerned with efficient use". [DETR 1998]

In order to make the operation of an ESCO attractive to the householder and compensate for the disruption of upgrading, the annual cost of energy plus repayments for the improvements would be, say, 25% less than previously paid for energy alone on a 'same warmth basis'. At the lower end of the SAP scale we could be talking of a 75 - 80%

improvement in the thermal efficiency of the property so that 'same warmth' energy bills would show an equivalent reduction, leaving a significant margin for increased comfort.

However, the incorporated upgrading repayments would fall considerably short of market rates of return. For public housing authorities this problem could be solved by resorting to public finance initiative (pfi) funding to bridge the repayments gap. It is the private sector which poses the most intractable problem.

If there is to be a significant participation in an upgrading programme by home owners there has to be a package of incentives. For many in this sector the cost of energy is not a major problem. Nevertheless, it is probably true that most owner occupiers would not sign up for refurbishing unless the combined cost of energy and upgrading came to less than the previous cost of energy alone. How do you establish that cost on an equitable basis across a range of properties and occupancy levels? There would have to be benchmark figures based on:

- the type of property as defined in the English House Condition Survey
- the scale and type of occupancy
- location and climate zone

The benchmarking would have to be undertaken by the local authority.

Another problem is that, with deregulation, consumers can change energy supplier at 28 days notice. An ESCO would require security of contract before embarking on work for a contract management body. One solution would be for the debt for the upgrading work to be transferred on completion to a third party. This would enable householders to exercise their rights under the 28 day rule. The ideal solution would be for the government to establish a refurbishment bank which would provide finance at below market rates, along the lines of the student loan scheme. This way the ESCO would not carry the financial burden of a neighbourhood programme but would be reimbursed on a house by house basis. At the same time the householders would gain a substantial improvement to their property reflected in an increase in market value.

The government is sympathetic to the principle that all houses should be energy rated at the point of sale. If and when this becomes mandatory and if energy prices start to rise driven by climate change pressures and supply uncertainties, then, during the initial stages of the programme, upgraded properties will have a clear market advantage. At the same time, improved properties will incur markedly reduced maintenance costs. Finally, we should not underestimate the importance of comfort gains, even for householders previously prepared to meet substantial energy costs to achieve what they perceive as adequate thermal comfort.

Private rented accommodation

Undoubtedly the most challenging problem concerns the private rented sector. How can landlords be attracted to the programme when it is their tenants who normally meet energy costs? There may be no way of avoiding coercive legislation. For example, if legislation were enacted which requires vendors to provide an energy rating of their homes at the point of sale, it would be reasonable to expect landlords to provide incoming tenants with a rating for their apartment. This would provide an incentive to upgrade the property and adjust the rent accordingly. This would have cost implications for the government in terms of increased housing benefit claims. Considering this is by far the most energy inefficient sector of the housing stock, this would not be an unreasonable burden for the government to shoulder.

From the landlord's point of view, the upgrading increases the stock value and the increased rent income strictly geared to the low interest repayments would mean that the upgrading costs were covered. There might have to be special arrangements for student housing, especially where rents are decided by the university/college at a level deemed to be affordable.

To be even more draconian, consideration should be given to a requirement that all accommodation offered for rent should achieve a minimum of SAP 60 and a target date set, say 2010, for compliance. In the meantime, there should be more rigorous application of the regulation that if a property is materially altered then the whole property should be raised to current Building Regulations standards.

What, then, is preventing these eminently reasonable arguments from carrying the day? Before there can be a national upgrading programme to achieve an overall minimum of SAP 60, a number of barriers will have to be removed. What are they?

The costs are considerable. It is estimated that the annual cost of upgrading all properties in England below SAP 50 to SAP 60 over a ten year programme would be £6.641 billion; over fifteen years: £4.408 billion. This should come as no surprise to the government since, some years ago, the RIBA advised a House of Lords Select Committee that the annual costs would be in the region of £4 billion over ten years but that these costs would be balanced by the cost of saved energy and this was stated in the Committee's report [HMSO 1991].

When benefits are factored in the picture becomes less daunting, even to justifying the claim that a ten year upgrading programme will produce a net benefit to the Exchequer. Wider benefits will be identified as the European Commission 'ExternE' project delivers results. This is a comprehensive attempt to use a consistent 'bottom-up' methodology to evaluate the external costs of different fuels.

Why, then, are we not tackling this problem head-on? What are the barriers?

So far, all the projects which meet or improve on the SAP 60 target have a captive clientele as in the case of Penwith. The success of a national rolling programme will depend on a significant participation rate within the private sector. This will only happen if there are seen to be adequate benefits which outweigh costs. The problem here is cheap energy which creates extended payback times for most energy efficiency measures. However, it is likely that we are currently 'enjoying' the quiet before the storm in the case of oil prices. According to the body 'Petroconsultants', oil production will reach its peak in two years time. "The world's oil companies are now finding only one barrel of oil for every four that we consume". [New Scientist 1999] A further disturbing fact is that about half the oil remaining is in the 'feudal' countries of the Middle East that can adjust the price according to market conditions. We saw the effect of this in 1973. There is a good chance that the era of cheap energy is coming to an end. Householders beware!

A second problem concerns the capability of the industry to provide the suitably qualified operatives for a crash programme of upgrading. This is a relatively specialised field and authorities undertaking upgrading programmes have encountered serious problems of quality control. When Sweden upgraded its thermal regulations under SBN 80, it gave two years notice and educational support to enable operatives to be equal to the more stringent construction standards. The UK should be doing this, now! Perhaps a level 3 NVQ for environmental construction technicians would be a start.

Thirdly there is the problem of delivery. It is generally accepted that the ESCO is the favoured delivery mechanism, but energy service companies of the size required to undertake neighbourhood scale projects are in short supply. Basically this is because the rates of return on retrofitting contracts which include the supply of energy are only about half the market rate. There will have to be government intervention to make ESCOs viable. In addition there will have to be rigorous management structures to impose quality and cost control. Local authorities may be the obvious holders of the contracts, but here too there would have to be a quantum jump in professional expertise. This was revealed by the wide variation in the quality of response to requirements laid on LAs under the Home Energy Conservation Act.

Fourth, there will need to be a fundamental change of stance on the part of the Energy Regulator, with energy efficiency taking precedence over energy price.

Finally, there is the problem of the way the Treasury does its sums. It is a classic case of tunnel vision that allows officials to see capital costs but makes them blind to revenue benefits. This, combined with short termism in expenditure allocations, makes the prospect of a capital intensive national upgrading programme inconceivable at a time of severe spending constraint. We are tyrannised by the Public Spending Borrowing Requirement. Politicians are laying great stress these days on the importance of' 'joined-up thinking'. Never has this message been more relevant than for the Treasury. If there is to be a breakthrough in tackling major problems like housing and renewable energy for the next century then capital costs must be considered in the context of revenue gains. Capital cost and revenue benefit are two sides of the same coin. An economic model which offsets costs with benefits will show that dealing with the housing problem in a systematic and fundamental way generates numerous interlocking virtuous circles. Here are some examples.

There will be a significant boost to the construction industry. One new skilled job is created for £30-35,000 spent on upgrading homes. In addition, there should be one new job for every £60-80,000 due to the respend factor. The support industries, particularly in the insulation and glazing sectors will see a significant increase in turnover. Construction is one of the highest unemployment sectors, so there should be an appreciable switch from unemployment benefit to tax and social security revenue. According to the Association for the Conservation of Energy [ACE 1998] the estimated number of jobs created each year by upgrading to SAP 60 would be:

Over 10 years   206638
Over 15 years   137758
Over 20 years   103319

These include direct and indirect jobs, including management.

Since this programme will involve all income groups, the cost of saved energy and reduced maintenance costs will amount to a significant increase in disposable income, even after some of the benefit has been taken in increased comfort. As energy prices are predicted to rise above inflation so the value of saved energy will increase, adding value to upgraded properties.

The national stock value of housing will be substantially increased, with the added benefit of longer life expectancy.

The improvements in health following upgrading will have economic benefits. Brenda Boardman has estimated that £1 billion per year is spent by the Health Service on illnesses directly attributable to cold and damp homes. This figure may be significantly higher in that it is impossible to quantify the contribution of poor housing to depressive illnesses. The DETR acknowledges that fuel poor households "also suffer from opportunity loss, caused by having to use a larger portion of income to keep warm than other households. This has adverse effects on the social well-being and overall quality of life for both individuals and communities". [DETR 1999] This cost will taper off as the upgrading programme gathers momentum.

What also remains to be assessed is the economic benefit of improved well-being on the part of those in employment. When the Lockheed Corporation commissioned a new headquarters building, it decided on an energy efficient design even though this added 4% to the costs. The environmental quality of the new building resulted in a drop in absenteeism of 15% which paid for the extra cost of construction in one year. This has obvious implications for healthier homes.

Finally, we should not overlook the health of the planet. Housing accounts for almost 29% of the total carbon dioxide emissions in the UK. A SAP 60 programme could ultimately result in a 50% reduction in the consumption of delivered energy within this sector even after the uptake of the improved comfort option. A contribution of 14.5% reduction in the UK emissions on the part of the housing sector would go a long way to enabling the government to meet its declared target of a 20% reduction against 1990 levels by 2010. At present the country is falling short of its lesser commitment under the Kyoto agreement for European Union countries.

Considerable publicity has been given to the need for 4.4 million new homes. It is politically much more attractive to talk of new build than retrofit projects. Not only is there the excitement of the new there is also the fact that new homes generate revenue without being a capital cost to the Exchequer. Existing housing should be given a much higher profile in the drive to improve the quality of life in the next millennium.

The drive to realise the regeneration of cities has been given added impetus by the report of Lord Rogers' Urban Task Force. [Urban Task Force 1999] This is a challenging and stimulating document which needs to be seen in the context of the 4.3 million households in England which fall below the official fuel poverty level. It is housing which underpins cities. If the underpinning is decrepit then city centre regeneration is built on insecure foundations. We have to see regeneration as a holistic problem, with the upgrading of homes taking place in parallel with the regeneration of central city areas. According to the Charity Shelter, in the mid 1970s Britain was near the top of the league as measured by the proportion of Gross Domestic Product spent on housing. By 1994 it had dropped to near the bottom of the table and things have not improved significantly since then.

Perhaps the most serious problem we face is that all the economic trends are in the direction of the free market meaning that all expenditure must be justified in terms of the market place. It is difficult to convince power brokers that, in the longer term, raising the quality of the housing stock will be cost effective by the most stringent of free market criteria. The process must be kick-started by the government investing in the longer term future, which, it seems, only governments are in a position to do. Surely a government which takes this step of faith will be seen by future generations as one of the beacons of social enlightenment.

Glossary

Affordable warmth - the ability to afford adequate energy services for 10% of income. The definition of income may or may not exclude housing costs, though it should - see Affordable Warmth Index (Chapter 8).

Deprivation indices (Townsend Index, Carstairs Deprivation Score, Jarman Score) – indices of deprivation set against certain indicators (e.g. Townsend uses unemployment, overcrowding, non-owner occupation and non-access to a car).

Dew point – The temperature at which water will begin to condense from a body of air (°C).

Energy Ratings (NHER, SAP) – A measure of the energy efficiency of a building. The oldest and most comprehensive is National Home Energy Rating (NHER) which is based upon the total annual running costs per square metre of dwelling. The Standard Assessment Procedure (SAP) is based solely on the space and water heating costs per square metre and is independent of location.

Excess Winter Deaths- the percentage by which the mortality rate for the period December to March exceeds that of other months of the year

Fuel poverty - the inability to afford adequate energy services, due to the energy inefficiency of the home.

ICD Codes - International Classification of Diseases, Injuries and Causes of Death. In the ICD coding system each disease/cause of death has an ICD number thus enabling consistency of definition internationally.

Ischaemic Heart Disease – Chronic disease of the heart which can lead to heart attacks

kWh - kilowatt hour – a measure of energy output - obtained my multiplying the energy rate in kilowatts (kW) by the number of hours.

Relative humidity – The ratio between the amount of moisture in a body of air and the maximum amount which it could contain at that temperature (%).

Seasonal mortality - seasonal fluctuation in deaths.

SF-36 – a health survey measuring health perceptions via 35 items measuring health across 8 dimensions and one item measuring health change, usually self-completed.

Standard heating regime –demand temperatures used in the BREDEM standard occupancy factors. These specify a demand temperature in the main living area of 21°C and 18°C in the rest of the property.

Standard occupancy pattern - a prescribed pattern of heating a dwelling, use of hot water and appliances and a standard number of people for a given dwelling area

Standardised Mortality Ratio -The local death rate as a percentage of the national rate, standardised for differences in age and sex composition.

U-value - the thermal transmission of a building element (heat flow per square meter surface area through the building element per degree temperature difference across the element) (w/m^2) – the lower the better for energy efficiency.

References

Numbers in square brackets give the chapter(s) where reference appears

Abramson M, Voight T (1991) Ambient air pollution and respiratory disease. *Med J Aust*: **154** 543-551 [5]

ACE (1998) *Green Job Creation in the UK,* Association for the Conservation of Energy, London [19]

Acheson D (1998) *Independent Inquiry into Inequalities in Health.* London: The Stationery Office [Pref, 12, 18]

Adan OCG (1994) *On the fungal defacement of interior finishes,* PhD thesis, Netherlands Organisation for Applied Scientific Research (TNO) [9]

Alderson MR (1985) Season and Mortality *Health Trends,* **17**, 87-96. [4]

Ambrose P (1996) *I Mustn't Laugh Too Much: Housing and Health on the Limehouse Fields and Ocean Estates in Stepney,* Centre for Urban and Regional Research, University of Sussex [10, 17, 18]

Ambrose P, Barlow J, Bonsey A, Pullin M, Donkin V and Randles J (1996) *The real cost of poor homes: A critical review of the research literature by the University of Sussex and the University of Westminster.* London: The Royal Institution of Chartered Surveyors. [11, 17, 18]

Ambrose P and Randles J (1999) *Looking for the Joins: A Qualitative Study of Inter-Agency Working in Holly Street and Hackney,* London Borough of Hackney [17]

Anderson BR, Clark AJ, Baldwin R and Millbank NO (1985) *BREDEM - BRE Domestic Energy Model: background, philosophy and description.* BRE Report, BRE, Watford. [8]

Anderson BR, Chapman PF, Cutland NG, Dickson CM and Shorrock LD (1996) *BREDEM-12 Model Description,* BRE Report BR315, BRE Watford [8]

Andrae S, Axleson O, Bjorksten B, Fredriksson M, Kjellman N-IM (1998) Symptoms of bronchial hyperreactivity and asthma in relation to environmental factors *Arch Dis Child*: **63** 473-478. [5]

Arblaster L and Hawtin M (1993) *'Health, Housing and Social Policy* Socialist health Association London [6]

Arlian L (1992) Water balance and humidity requirements of house dust mites, *Exp. & Applied Acarology,* **16,** pp 15-35 [9]

Arlian LG (1989) Biology and ecology of house dust mites *Dermatophagoides* spp. and *Euroglyphus* spp. *Immunol Allergy Clin N Amer*: **9** 339-356. [5]

Arnstein SR (1969) A ladder of citizen participation *AIP J.* July 1969 pp 216-224 [13]

Ashley J, Smith T, Dunnell K (1991) Deaths in Great Britain associated with the influenza epidemic of 1989/90. *Population Trends* **62:**16-20 [2]

Ashmore I (1998) Asthma, housing and environmental health, *Environ Health,* January 1998, pp 17-23 [9]

Aspinall PJ (1994) *Health and housing in Bromley part 1 - Housing and health: An overview.* Tunbridge Wells: South East Institute of Public Health. [18]

Association for the Conservation of Energy (ACE) and Projects in Partnership (PIP), (1999) *Domestic Energy Efficiency and Health*, A report to the Eaga Charitable Trust [16]

Association of Metropolitan Authorities (1997) *Housing and health: Getting it together.* London: Association of Metropolitan Authorities (obtainable via Local Government Association). [18]

Åstrand P-O, Ekblom B, Messin R, Saltin B and Stenberg J (1965) Intra-arterial blood pressure during exercise with different muscle groups. *J. Appl. Physiology*, **20**, 253. [4]

Austin JB, Russell G, Adam MG, Mackintosh D, Kelsey S, Peck DF (1994) Prevalence of asthma and wheeze in the Highlands of Scotland. *Arch Dis Child*: **71** 211-216. [5]

Aylin P (1999) *Excess winter mortality, housing and socio-economic deprivation.* Paper presented at the Symposium on Health, Housing and Affordable Warmth, London [11]

Bainton D , Moore F and Sweetnam P (1977) Temperature and deaths from ischaemic heart disease. *British J. of Preventative and Social Medicine*, **31**, 49-53. [4]

Bardsley M, Rees Jones I, Kemp V, Aspinall P, Dodhia H, Bevan P (1998) *Housing and health in London*. The Health of Londoners Project, Directorate of Public Health, East London and The City Health Authority [12, 18]

Barnes PJ, Rodger IW, Thomson NC (Editors) (1992) *Asthma - Basic Mechanisms & Clinical Management - Second Edition*. Academic Press Ltd. London [5]

Barrow M and Bachan R (1997) *The real cost of poor homes: Footing the bill.* London: The Royal Institution of Chartered Surveyors [11, 17, 18]

Baumgart P (1992) Impact of shifted sleeping and working phases on diurnal blood pressure rhythm (in) *Temporal Variations of the Cardiovascular System* (Eds: Schmidt, Engel and Blümchen) Berlin: Springer-Verlag, 283-296. [4]

Bean WB and Mills CA (1938) Coronary occlusion, heart failure and environmental temperatures. *Am Heart J*, **16**, 701. [4]

Becker RC and Carrao JM (1989) Circadian variations in cardiovascular disease. *Cleveland Clinical J. of Medicine*, **56**, 676-680. [4]

Bexley & Greenwich Health Authority (1996) *Annual Public Health Report 1996.* London: Bexley & Greenwich Health Authority. [18]

BGAM (1997) The British Guidelines on Asthma Management 1995 review and position statement. *Thorax* **52(1)**:S2-21 [12]

Birmingham City Council (1997) *Health and Housing Conditions and the Impact of Poverty*, joint report of the Director of Housing and the Director of Environmental Services. [15]

Birmingham City Council (1998) *Sparkbrook, Sparkhill and Tyseley Regeneration Area Urban Care Needs Survey*. [15]

Birmingham Health Authority (1995) *Closing the Gap,* Birmingham Public Health Report. [15]

Birmingham Health Authority (1999) *Birmingham's Health Improvement Programme, Working Together for the Health of Local People.* [15]

Blythe ME (1976) Some aspects of the ecological study of the house dust mite. *Brit J Dis Chest* **70** 3-21. [5]

Boardman B (1991) *Fuel Poverty: from cold homes to affordable warmth.* Belhaven Press London [Pref, 2, 4, 6, 7, 10, 17, 18]

Boardman, B (1993) in *Proceedings of Neighbourhood Energy Action Conference, Birmingham,* National Energy Action (NEA) Newcastle upon Tyne. [10]

Bordass W and Oreszczyn T, (1998) *Internal Environments in Historic Buildings: Monitoring, Diagnosis and Modelling,* Report to English Heritage, October 1998 [9]

Bowling, A (1991) *Measuring Health,* Oxford University Press, Oxford [17]

Boyd D, Cooper P, Oreszczyn T, (1988) Condensation risk prediction: Addition of a condensation model to BREDEM, *Building Serv Eng Res & Technol,* **9**(3), pp 117-125, [9]

Brazier JE, Harper R, Jones NMB, O'Cathain A, Thomas KJ, Usherwood T and Westlake L (1992) Validating the SF – 36 health survey questionnaire: a new outcome measure for primary care. *Brit Med. J.* **305**: 160-4 [7]

Brazier JE, Howard P and Williams BT MRC Report: *'Comparison of Outcome measures for patients' with COPD.'* Grant no: *G9203850.* Sheffield: SCAAR, 1994. [7]

Brennan PJ, Greenberg G, Miall WE, Thompson SG (1982) Seasonal variation in arterial blood pressure. *Brit. Med. J.,* **285**, 919-923. [4]

British Medical Association (1998) *Health and environmental impact assessment.* London Earthscan Publications Ltd. [18]

Bronswijk J (1981) *House Dust Biology,* Zoelmond, Netherlands [9]

Brunekreef B (1992) Damp housing and adult respiratory symptoms. *Allergy* 1992; **47** pp498-502 [12]

Brunekreef B, Dockery DW, Speizer FE, Ware JH, Spengler JD, Ferris BG (1989) Home Dampness and respiratory morbidity in children. *Am Rev Respir Dis*: **140** 1363-1367. [5]

BS5250 (1989) *British Standard Code of practice for control of condensation in buildings,* British Standard Institute. [9]

Buckland FC and Tyrrell DAJ (1962) Loss of infectivity on drying various tissues. *Nature,* **195**, 1063-4. [3]

Bucknall Austin (1994) *Birmingham Private Sector House Conditions Survey.* [15]

Building Research Establishment (May 1985) *BRE Digest 297: Surface Condensation and mould growth in traditionally-built dwellings.* BRE, Garston. [7]

Bull GM (1973) Meteorological correlates with myocardial and cerebral infarction and respiratory disease. *British J. of Preventative and Social Medicine,* **27**, 108-113. [4]

Bull GM and Morton J (1975) Relationships of Temperature with death rates from all causes and from certain respiratory and arteriosclerotic diseases in different age groups. *Age and Ageing,* **4**, 232-246. [4]

Bull GM and Morton J (1978) Environment, temperature and death rates. *Age, ageing,* 7: 210-21. [1]

Burgess I (1993) Allergic reaction to anthropods. *Indoor Environ* **2** 64-70. [5]

Burr ML, Butland BK, King S, Vaughan-Williams E (1989) Changes in asthma prevalence: two surveys 15 years apart. *Arch Dis Child*: **64** 1452-1456. [5]

Burr ML, Miskelly FG, Butland BK, Merrett TG, Vaughan-Williams E (1989) Environmental factors and symptoms in infants at high risk of allergy. *J Epidemiology Community Health:* **108** 99-101. [5]

Burr ML, St Leger AS, Yarnell JWG (1981) Wheezing, dampness and coal fires. *Community Med*: **3**: 203-209. [5]

Burridge R and Ormandy D (Eds.) (1993) *Unhealthy Housing: Research, Remedies and Reform,* E and FN Spon, London [6, 17, 18]

Burrows and Walentowicz (1992) *Homes cost less than homelessness.* London: Shelter. [18]

Carr-Hill R (1997) *The impact of housing conditions on health status.* Unpublished paper delivered at 'The wider impact of housing' conference, Scottish Homes, April [18]

Carr-Hill R, Coyle D and Ivens C (1993) *Poor housing: poor health?* Unpublished report funded by the Department of the Environment. [11, 17, 18]

Carstairs V, Morris R (1991) *Deprivation and health in Scotland.* Aberdeen, Aberdeen University Press. [2]

Carswell F, Birmingham K, Oliver J, Crewes A, Weeks J (1996) The respiratory effects of reduction of mite allergen in the bedrooms of asthmatic children – a double-blind controlled trial. *Clin Exp Allergy* 1996 **26(4)** 386 – 396. [5]

Carswell F, Robinson DW, Oliver J, Clark J, Robinson P, Wadsworth J (1982) House Dust Mites in Bristol. *Clin Allergy* **12** 533-545. [5]

Central Policy Unit (1993) Sheffield City Council [7]

Chantler C and Kelly S (1999) Deaths from Hypothermia in England and Wales. Health Statistics Quarterly Summer 1999 p 50--51 [18]

Chapman, J (1991) The development of the National Home Energy Rating. *Proc BEPAC Conference. Building Environment Performance '91* BEPAC [8]

Choquette G and Ferguson RJ (1973) Blood pressure reduction in borderline hypertensives following physical training. *Canadian Medical Association J.,* **108**, 699-703. [4]

CISHA (1994) *Health of the Population 1994,* Annual Report of the Director of Public Health Medicine, Cornwall and Isles of Scilly Health Authority, St. Austell, Cornwall [12]

Cole I and Furbey R (1994) *The Eclipse of Council Housing,* Routledge, London [7]

Coleshaw SRK, Syndercombe-Court D, Donaldson GC, Easton JC, Keatinge WR (1990) Raised platelet count and RBC in human subjects during brief exposure to cold and active rewarming. *Archives Complex Environmental Studies,* **2**,15-24. [4]

Collins, KJ (1986) Low indoor temperatures and morbidity in the elderly. *Age and Ageing,* **15**, 212-20. [3, 4]

Collins KJ (1987) Effects of cold on elderly people. *British J. of Hospital Medicine,* **38**, 6, 506-514. [4]

Collins, KJ (1993) Cold- and heat-related illnesses in the indoor environment, in *Unhealthy Housing: Research, Remedies and Reform* (Eds Burridge R and Ormandy D) E & FN Spon, London, 117-40. [3]

Collins, KJ (1995) Hygrothermal conditions. *In Building regulation and health* (Eds GJ Raw and RM Hamilton), Building Research Establishment, Watford [11]

Collins KJ, Abdel-Rahman TA, Easton JC, Sacco P, Ison J and Doré CJ (1996) Effects of facial cooling on elderly and young subjects: interactions with breath-holding and lower body negative pressure. *Clinical Science,* **90**,485-492. [4]

Collins KJ, Abdel-Rahman TA, Goodwin J and McTiffin L (1995) Circadian body temperatures and the effects of a cold stress in elderly and young subjects. *Age and Ageing,* **24**, 485-489. [4]

Collins KJ, Easton JC, Belfield-Smith H , Exton-Smith AN and Pluck RA (1985) Effects of age on body temperature and blood pressure in cold environments. *Clinical Science,* **69**, 465-470. [4]

Collins KJ and Goodwin J (1998) Cardiovascular responses of the elderly in cold environments. Yokohama, Japan: National University Press, *Proc 2nd Int Conf Hum Env System,* 307-311 [4]

Collins KJ and Hoinville E (1980) Temperature requirements in old age. *Building Services Engineering Research and Technology,* **1**, 165-72. [3, 4]

Collins KJ, Sacco P, Easton JC, Abdel-Rahman TA (1989) Cold pressor and trigeminal cardiovascular reflexes in old age (in) *Thermal Physiology* (Ed: JB Mercer). London: Elsevier Science Publishers. [4]

Colloff MJ (1986) The use of liquid nitrogen in the control of house dust mite population. *Clin Allergy* **16** 42-47. [5]

Colloff MJ (1990) House dust mites-part II. Chemical control. *Pestic outlook* **1**(2) 3-8. [5]

Colloff MJ, Ayres J, Carswell F, Howarth PH, Merrett TG, Mitchell EB, Walshaw MJ, Warner JO, Warner JA, Woodcock AA (1992) The control of allergens of dust mites and domestic pets: a position paper. *Clin Exp Allergy* **22**(2) 1-28. [5]

Colloff MJ, Taylor C, Merrett TG (1995) The use of domestic steam cleaning for the control of house dust mites. *Clin Exp Allergy* **25** 1061-1066. [5]

Crombie DL, Fleming DM, Cross KW, Lancashire RJ (1995) Concurrence of monthly variations of mortality related to underlying cause in Europe. *J Epidemiol Community Health;* **49**(4):373-378 [2]

Croydon Health Think-Tank (1995) *Final report from Croydon North Health Think-Tank / Final report from Croydon South-East Health Think-Tank.* Croydon: London Borough of Croydon and Croydon Health Authority. [18]

Cuijpers CEJ, Swaen GMH, Wesseling G, Sturmans F, Wouters EFM (1995) Adverse effects of the indoor environment on respiratory health in primary school children. *Environmental Research* **68** pp11-23 [12]

Cunningham MJ (1996) Controlling Dust Mites Psychrometrically - A Review for Building Scientists, *Indoor Air*, pp 249-258 [9]

Curwen M (1991) Excess winter mortality: a British phenomenon? *Health Trends* **22(4):**169-75 [2, 4]

Curwen M (1997) Excess winter mortality in England and Wales with special reference to the effects of temperature and influenza (in) *The Health of Adult Britain Vol I and II (Eds Charlton J and Murphy M)* Dicentennial Suppl No **12**. London: The Stationery Office. [3, 4, 10, 18]

Curwen M and Devis T (1988) Winter mortality, temperature and influenza: has the relationship changed in recent years? *Population Trends* **54**, 17-20 [11]

Dales RE, Zwanenburg R, Burnett R, Franklin CA (1991) Respiratory health effects of home dampness and molds among Canadian children. *Am J Epidemiol*; **134** pp196-203 [12]

Dannenberg AL, Keller JB, Wilson PWF, Castelli WP (1989) Leisure time physical activity in the Framingham Offspring Study: description, seasonal variation and risk factor correlates *American J. of Epidemiology*, **129**, 1, 76-88. [4]

Dekker C, Dales R, Bartlett S, Brunekeef B, Zwanenburg H (1991) Childhood asthma and the indoor environment. *Chest* **100** 922-926. [5]

Dekker H (1928) Asthma und milben. *Munch Med Wochenschr* **75** 515-516. [5]

Department of Health (1996) *Health Service Indicators.* London: Department of Health. [18]

Department of Health (1997) *Public Health Common Data Set 1997.* London: Department of Health. [18]

Department of Health (1998) *The new NHS: modern, dependable.* Department of Health, The Stationery Office, London, 1998 [12, 18]

Department of Health (1998) *Our Healthier Nation; A Contract for Health*, The Stationery Office, London [16]

Department of Health (1998b) *Partnership in Action: New opportunities for joint working between health and social services – A discussion document*, The Stationery Office, London. [16]

Department of Health (1999) *Saving Lives. Our Healthier Nation.* Department of Health, The Stationery Office, London. [12, 18]

Department of Health, (1999) press release 9/2/99: *Partnership is at the centre of the new Health Bill*, Department of Health, London [16]

Department of the Environment (DoE) (1993) *English House Condition Survey 1991.* London: HMSO. [3, 18]

Department of the Environment (DoE) (1996) *English House Condition Survey 1991: Energy Report* HMSO, London [Pref, 2, 3, 4, 9, 10, 11, 15, 19]

Department of the Environment, Energy Efficiency Office (1994) *The Government's Standard Assessment procedure for energy rating of dwellings.* BRESCU, Building Research Establishment, Garston. [2]

Department of the Environment Transport and the Regions (1997) *Private Sector Housing Renewal Strategies, Good Practice Guide.* HMSO [15]

Department of the Environment Transport and the Regions (DETR) (1998) *Unlocking the potential - financing energy efficiency in private housing,* DETR General Information Report 50, The Stationery Office [19]

Department of the Environment Transport and the Regions (1998) *Single Regeneration Budget bidding guidance.* The Stationery Office London: [18]

Department of the Environment Transport and the Regions (1998) *Modern Local Government. In touch with the people,* The Stationery Office, London. [16]

Department of the Environment Transport and the Regions (DETR) (1999) *New Deal for Communities. Developing delivery plans* The Stationery Office [13]

Department of the Environment Transport and the Regions (1999) *Fuel Poverty: the New HEES — a programme for warmer, healthier homes. London,* DETR. [Pref, 2,10, 19]

Department of Trade & Industry (1995) *Home Accident Surveillance System Annual Report 1995.* London: Department of Trade & Industry. [18]

Dickson CM, Dunster JE, Lafferty SZ and Shorrock LD (1996) BREDEM: Testing monthly and seasonal versions against measurements and against detailed simulation models. *Building Serv.Eng. Res. Technol* **17**(3) pp 135-140 [8]

Donaldson GC and Keatinge WR (1997) 'Mortality related to cold weather in elderly people in South East England 1970-1994'. *Brit. Med. J.,* **315**: 1055-6. [1, 18]

Donaldson GC, Robinson D and Allaway SL (1997) An analysis of arterial disease mortality and BUPA health screening in men, in relation to outdoor temperature. *Clinical Science,* **92,** 261-268. [1, 4]

Donaldson GC, Keatinge WR (1997) Early increases in ischaemic heart disease mortality dissociated from, and later changes associated with, respiratory mortality, after cold weather in south east England. *J. Epidem. Comm. Health,* **51**: 643-8. [1]

Donaldson GC, Tchernjavskii VE, Ermakov SP, Bucher K, Keatinge WR (1998a) Winter mortality and cold stress in Yekaterinburg, Russia: interview survey. *Brit. Med. J.,* **316,** 514-518. [4]

Donaldson GC, Ermakov SP, Komarov YM, McDonald CP, Keatinge WR (1998b) Cold related mortalities and protection against cold in Yakutsk, eastern Siberia: observation and interview study. *Brit. Med. J.,* **317,** 978-982. [1, 3, 4]

Donaldson GC, Tchernjavskii VE, Ermakov SP, Bucher K, Keatinge WR (1998c) Effective protection against moderate cold, with rise in mortality only below 0 °C, in Yekaterinburg, Russian Federation. *Brit. Med. J.,* **316**: 514-8. [1]

Dorman PJ, Slattery J, Farrell B, Dennis MS, Sandercock PAG (1997) A randomised comparison of the Euroqol and Short Form-36 after stroke. *Brit. Med J* **315** 461. [5]

Dorward AJ, Colloff MJ, MacKay NS, McSharry C, Thomson NC (1988) Effect of house dust mite avoidance measures on adult atopic asthma. *Thorax* **43** 98-102. [5]

Douglas AE, Hart BJ (1989) The significance of the fungus *Aspergillus penicilloides* to the house dust mite *Dermatophagoides pteronyssinus*. *Symbiosis* **7** 105-117. [5]

Dowse GK, Turner KJ, Stewart GA, Alpers MP, Woolcock AJ (1985) The association betwen *Dermatophagoides* mites and the increasing prevalence of asthma in village communities within the Papau New Gineau Highlands. *J Allergy Clin Immunol* **75** 75-83. [5]

Drennan V and Tearn J (1986) 'Health visitors and homeless families'. *Health Visitor,* **59**: 340-2. [18]

Druett, HA (1967) The inhalation and retention of particles in the human respiratory system: airborne microbes, in *17th Symposium of the Society for General Microbiology* (eds PH Gregory and JL Monteith), Cambridge University Press, Cambridge. [3]

Duke PC, Wade JG, Hickey RF and Larson CP (1976) The effects of baroreceptor reflex function in man. *Canadian Anaesthetic Society J.,* **2**, 111-124. [4]

Dunnigan MG, Harland WA and Fyfe T (1970) Seasonal incidence and mortality of ischaemic heart disease. *Lancet, 2, 793-797.* [4]

EAGA (1999) *Housing and Health; The Cornwall Intervention Study (unpublished seminar contribution)*, EAGA Charitable Trust, [19]

East London & The City Health Authority (1998) *Health in the East End: Annual Public Health Report 1989/99.* London: East London & The City Health Authority. [18]

Easterlow D and Smith SJ (1997) 'Fit for the future? A role for health professionals in housing management'. *Public Health,* **111**: 171-8. [18]

Eckberg DL and Sleight P (1992) *Human Baroreflexes in Health and Disease.* Oxford: Clarendon Press. [4]

Ehnert B, Lau-Schadendorf S, Weber A, Buettner P, Schou C, Wahn U (1992) Reducing domestic exposure to dust mite allergens reduces bronchial hyperreactivity in sensitive children with asthma. *J Allergy Clin Immunol* **90** 135 – 138. [5]

Elton PJ and Packer JM (1986) 'A prospective randomised trial of the values of re-housing on the grounds of mental ill health'. *J. of Chronic Disease,* **39**: 221. [18]

Enquselassie F, Dobson AJ, Alexander TM, Steele PL (1993) Seasons, temperature and coronary disease. *Internatl J Epidemiol* **22(4)**:632-636 [2]

Erikssen J and Rodahl K (1979) Seasonal variation in work performance and heart rate response to exercise. A study of 1835 middle aged men. *European J. of Applied Physiology,* **42**, 133-140. [4]

Eurowinter Group: Keatinge WR, Donaldson GC, Bucher K, Cordioli E, Dardanoni L, Jendritzky G, Katsouyanni K, Kunst AE, Mackenbach JP, Martinelli M, McDonald C, Näyhä S, Vuori I (1997) Cold exposure and winter mortality from ischaemic heart disease, cerebrovascular disease, respiratory disease, and all causes in warm and cold regions of Europe. *The Lancet,* **349**: 1341-1346. [1, 2, 3, 4]

Fleming DM, Cross KW, Crombie DL, Lancashire RJ (1993) Respiratory illness and mortality in England and Wales. *European J Epidemiol* **9(6):**571-576 [2]

Fox RH, Woodward PM, Exton-Smith AN, Green MF, Donnison DV and Wicks MH (1973) Body temperatures in the elderly: a National study of physiological, social and environmental conditions. *Brit. Med. J.,* **1**, 200-206. [4]

Friedewald VE and Spence DW (1990) Sudden cardiac death associated with exercise: the risk benefit issue. *American J. of Cardiology,* **66**, 183-188. [4]

Garrat J and Nowak (1991) *Tackling Condensation: The guide to the causes and remedies for surface condensation and mould in traditional housing,* Building Research Establishment, Garston. [7]

Garratt AM, Ruta DA, Abdulla MI, *et al.* (1993) The SF-36 health survey questionnaire: an outcome measure suitable for routine use within the NHS. *Brit. Med. J.* 306: 1440-4 [7]

General Register Office (1841) *Third Annual Report of Births Marriages and Deaths in England.* London: HMSO, 102-109. [4]

Giacone S, Ghione S, Palombo C, Genovesi-Ebert A, Marabotti C, Fommei E and Donato L (1989) Seasonal influences on blood pressure in high normal to mild hypertensive range. *Hypertension,* **14**, 22-27. [4]

Giaconi S and Ghione S (1992) Seasonal and Environmental Temperature Effects on Arterial Blood Pressure (in) *Temporal Variations of the Cardiovascular System* (Eds: Schmidt, Engel and Blümchen). Berlin: Springer-Verlag. [4]

Giesbrecht GG (1995) The respiratory system in a cold environment. *Aviation, Space and Environmental Medicine,* **66**, 890-902. [1, 3]

Glover GR (1996) 'Mental Illness Needs Index (MINI)' in Thornicroft G and Strathdee G (Eds). *Commissioning mental health services.* London: HMSO. [18]

Goldberg *et al.* (1990) 'The influence of social factors on common mental disorders'. *British J. of Psychiatry,* **156**: 704-13. [18]

Golding AMB (1987) 'Housing and respiratory disease'. *Public Health,* **101**: 317. [18]

Goodwin J (1999) *Seasonal cold, blood pressure and physical activity in young and elderly subjects.* PhD Thesis. University of Exeter. [4]

Gotzsche PC, Hammarquist C, Burr M (1998) House dust mite control measures in the management of asthma: meta-analysis. *Brit. Med. J.* **317** 1105 – 1109. [5]

Gravesen S (1979) Fungi as a cause of allergic disease. *Allergy* **34** 135-154. [5]

Green G and Gilbertson J (1999) Housing, Poverty and Health: The impact of housing investment on the health and quality of life of low income residents' *Open House International* **24**(1).[7]

Green G, Gilbertson J and Grimsley (Forthcoming) *Fear of Crime and Health in Residential Tower Blocks: A case study in Liverppool,* UK. [7]

Green, GH (1982) The positive and negative effects of building humidification. *ASHRAE Transactions,* **88**, Part 1. [3]

Greener M (1997) *The Which? Guide to Managing Asthma*. Which? Ltd, London. [5]

Griffiths S (1994) *Poverty on your doorstep: London Borough of Newham poverty profile*, Anti-Poverty and Welfare Rights Unit, L.B. of Newham. [10]

Hallas TE (1990) The biology of mites. *Allergy* **11** 6-9. [5]

Harshfield GA (1992) Factors associated with differences in the diurnal variation of blood pressure in humans (in) *Temporal Variations of the Cardiovascular System* (Eds: Schmidt, Engel and Blümchen), 272-282. Berlin: Springer-Verlag. [4]

Hart BJ, Whitehead L (1990) Ecology of house dust mites in Oxfordshire. *Clin Exp Allergy* **20** 203-209. [5]

Harving H, Korsgaard J, Dahl R (1988) Mechanical ventilation in dwellings as preventive measures in mite asthma. *Allergy Proc* **9** 283. [5]

Hata T, Ogihara T, Maruyama A, Mikami H, Nakamaru M, Naka T, Kumahara Y, Nugent CA (1982) The seasonal variation of blood pressure in patients with essential hypertension. *Clinical and Experimental Hypertension*, **4**, 341-354. [4]

Home Energy Conservation Act (HECA) (1995) *Department of the Environment: Circular 1996/2* HMSO (publ 1996) [10]

Heller RF, Rose G, Tunstall-Pedoe HD, Christie DGS (1978) Blood Pressure measurement in the UK Heart Disease Prevention Project. *J. of Epidemiology and Community Health*, **32**, 235-238. [4]

Henderson G and Shorrock LD (1986) BREDEM, the BRE Domestic Energy Model: Testing the predictions of the two zone model. *Building Serv. Eng.Res Technol.* 7(**2**) pp 87-91 [8]

Henwood M (1997) *Fuel Poverty, Energy Efficiency and Health*, A report to the Eaga Charitable Trust. [16]

Herity B, Daly L, Bourke GJ, Horgan JM (1991) Hypothermia and mortality and morbidity. *J. of Epidemiology and Community Health*, **45**, 19-23. [4]

Hewett C (1996) *Health and the Environment* NEA London [6]

Hill L (1928) The ciliary movement of the trachea studied in vitro: a measure of toxicity. *The Lancet*, **215**, 802-5. [3]

Hill S and Barlow J (1995) Single Regeneration Budget: Hope for those 'Inner Cities'? *Housing Review*, 1995: 44.2. 32-35 [17]

HMSO (1991) *Energy and the Environment*, House of Lords 13th report from the Select Committee of the European Communities, HMSO, [19]

Hopton JL and Hunt SM (1996) Housing conditions and mental health in a disadvantaged area in Scotland. *J. of Epidemiology and Community Health*, **50**, 56-61 [11]

HSC (1998) *Health Improvement Programmes: planning for better health*. Health Services Circular (HSC) 1998/167, Local Authorities Circular (LAC) *(98)23*. [12]

Humfrey C *et al.*, (1996) *Indoor Air Quality in the Home*, MRC Institute for Environmental Health Assessment A2 [9]

Hunt DRG and Gidman MI (1982) A National Field Survey of House Temperatures. *Building Environment*, **17**, 2, 107-124. [4]

Hunt S (1988) *Damp Housing, Mould Growth and Health Status* Part 1 - Report to the Funding Bodies [6]

Hunt S (1993) Damp and mouldy housing: a holistic approach? in *Unhealthy housing: research, remedies and reform, (Eds. Burridge R and Ormandy D)*, E&FN Spon, London pp.69-93 [7, 11]

Hunt S (1997) 'Housing and health' in Charlton J and Murphy M (Eds) *The health of adult Britain 1841-1994*. London: The Stationery Office. [18]

Hunt S and Boardman B (1994) 'Defining the problem' in *Domestic energy and affordable warmth*, (Ed.) T A Markus. Watt Committee on Energy, Report 30. Spon. London [Pref]

Hurley BF, Hagberg JM, Goldberg AP, Seals DR, Ehsani AA, Brennan RE and Holloszy JO (1988) Resistive training can reduce coronary risk factors without altering VO₂ max or percent body fat. *Medicine and Science in Sports and Exercise*, **20,** 2, 150-154. [4]

Huss K, Squire EN Jr, Carpenter GB, Smith LJ, Huss RW, Salata K, Salerno M, Agostinelli D, Hershey J (1992) Effective education of adults with asthma who are allergic to dust mites. *J Allergy Clin Immunol* **89(4)** 836 – 843. [5]

Hyndeman SJ (1990) 'Housing dampness and health among British Bengalis in East London'. *Social Science and Medicine,* **30**: 131-41. [3, 6, 11, 18]

Hyndeman SJ, Williams DR, Merrill SL, Lipscombe JM and Palmer CR (1994) 'Rates of admission to hospital for asthma'. *Brit. Med. J.,* **308**: 1506-600. [18]

Ineichen B (1993) *Homes and health: How housing and health interact.* London: E&FN Spon (Chapman and Hall). [18]

International Energy Agency (IEA) (1991) *Annex 14: Condensation and Energy: Volume 1: Sourcebook* [9]

Jaakola JJ and Heinonen OP (1995) Shared office space and the risk of the common cold. *European J. of Epidemiology*, **11**, 213-6. [3]

Jenner AD, English DR, Vandongen R, Beilin LJ, Armstrong BK, Dunbar D (1987) Environmental temperature and blood pressure in 9 year old Australian children. *J. of Hypertension*, **5**, 683-686. [4]

Joseph Rowntree Foundation (1996) *Local Maintenance Initiatives for Home Owners, Good Practice for Local Authorities.* [15]

Kammerling RM and O'Connor S (1993) 'Unemployment rates as a predictor of psychiatric admission'. *Brit. Med. J.,* **307**: 1536-9. [18]

Kannel WB, Wilson P, Blair SN (1985) Epidemiological assessment of the role of physical activity and fitness in development of cardiovascular disease. *American Heart J.*, **109**, 876-885. [4]

Keatinge WR (1986) Seasonal mortality among elderly people with unrestricted home heating. *Brit. Med. J.,* **293**, 732-733. [1, 4]

Keatinge WR, Coleshaw SRK, Cotter F, Mattock M, Murphy M and Chelliah R (1984) Increase in platelet and red cell counts, blood viscosity and arterial pressure during mild surface cooling: factors in mortality from coronary and cerebral thrombosis in winter. *Brit. Med. J.*, **289**, 1405-1408. [1, 2, 4]

Keatinge WR, Coleshaw SRK and Holmes J (1989) Changes in seasonal mortality with improvement in home heating in England and Wales 1964-1984, *International J. of Biometeorology,* **33**, 71-76. [1, 2, 3, 4]

Kellett JM (1993) Crowding and mortality in London Boroughs, in *Unhealthy Housing: Research, Remedies and Reform* (Eds. Burridge R and Ormandy D) E & FN Spon, London, 209-22. [3]

Khaw KT, Woodhouse P, Bulpitt CJ (1995) Interrelation of vitamin C, infection, haemostatic factors and cardiovascular disease. *Brit Med J.* **310**:1548-49,1559-63 [1, 2]

Khaw KT (1995) Temperature and cardiovascular mortality. *Lancet* **345**:337-338 [2, 4]

Knox EG (1981) Meteorological associations of cerebrovascular disease mortality in England and Wales. *J Epidemiol Community Health;* **35(3)**:220-223 [2]

Knuiman JT, Hautvast JGAJ, Zwiauer KFM (1988) Blood pressure and excretion of sodium, potassium, calcium and magnesium in 8 and 9 year old boys from 19 European centres. *European J. of Clinical Nutrition,* **42**, 847-855. [4]

Korkushko OV, Shatilo VB, Plachinda Y and Shatilo TV (1991) Autonomic control of cardiac chronotropic function in man as a function of age: assessment by power spectral analysis of heart rate variability. *J. of Autonomic Nervous System,* **32**, 191-198. [4]

Korsgaard J (1983) House-dust mites and absolute indoor humidity, *Allergy,* **38,** pp 86-96 [9]

Korsgaard J (1983) Preventive measures in mite asthma. A controlled trial. *Allergy* **38** 98-102. [5]

Korsgaard J (1983) Mite asthma and residency. *Am Rev Respir Dis* **128** 231-235. [5]

Korsgaard J and Iversen M (1991) Epidemiology of house dust mite allergy. *Allergy* **46(11)** 14-18. [5]

Kort HSM and Kneist FM (1994) Four-year stability of Der PI in house dust under simulated domestic conditions *in vitro. Allergy* **49** 131-133. [5]

Koselka H and Tukiainen H (1995) Facial cooling but not nasal breathing cold air induces broncho-constriction: a study of asthmatic and healthy subjects. *European Respiratory J.,* **8**, 2083-93. [3]

KPMG (1998) *What price energy efficiency;* summary of the KPMG Report, Energy Saving Trust, [19]

Kunst AE, Looman CWN, Mackenbach JP (1993) Outdoor air temperature and mortality in the Netherlands: a time-series analysis. *Am J Epidemiol* **137(3)**:331- 341 [2, 10]

Laake K and Sverre JM (1996) Winter excess mortality: a comparison between Norway and England plus Wales. *Age and Ageing* **25(5)**:343-348 [2, 4]

Lamb KL and Brodie DA (1990) The assessment of physical activity by leisure time physical activity questionnaires. *Sports Medicine,* **10**, 3, 159-180. [4]

Langham S, Normand C, Piercy J and Rose G (1994) 'Coronary heart disease' in Stevens A and Rafferty J (Eds). *Health Care Needs Assessment Volume I.* Oxford: Radcliffe Medical Press. [18]

Larsson K, Tornling G, Gavhed D *et al.* (1998) Inhalation of cold air increases the number of inflammatory cells in the lungs of healthy subjects. *European Respiratory J.*, **12**, 825-30. [3]

Latvala JJ, Reijula KE, Clifford PS *et al.* (1995) Cold-induced responses in the upper respiratory tract. *Arctic Medical Research,* **54**, 4-9. [3]

Lawson R (1997) *Bills of Health,* Radcliffe Medical Press, Oxford [11(66-87), 17]

Leather P, Mackintosh S and Rolfe S (1994) *Papering over the cracks: Housing conditions and the nation's health.* London: National Housing Forum. [18]

Leen MG, O'Connor T, Kelleher C, Mitchell EB, Loftus BG (1994) Home environment and childhood asthma. *Irish Medical J.* **87** pp142-4 [12]

Lindfors A, Wickman M, Hedlin G, Pershagen G, Rietz H, Nordvall SL (1995) Indoor environmental risk factors in young asthmatics: a case-control study. *Archives of Disease in Childhood* **73** pp408-12 [12]

London Accident Analysis Unit (1997) *Annual Report 1997.* London: London Research Centre. [18]

Lowry S (1989) Housing and health: Temperature and humidity. *Brit. Med. J* **299** 1326-1328. [5]

Lowry S (1991) *Housing and Health.* BMJ publications [6, 12]

Mancia G, Ferrari A, Gregorini L (1980) Blood pressure variability in man: its relation to high blood pressure, age and baroreflex sensitivity. *Clinical Science,* **59**, 401s-404s. [4]

Mansoor GA, McCabe EJ, White WB (1994) Long term reproducibility of ambulatory blood pressure. *J. of Hypertension,* **12**, 6, 703-708. [4]

Marchant M, Moore K and Ditchburn J (1997) *Hospitalisation due to injury: Lambeth, Southwark & Lewisham & South Thames (East) residents from 1989 to 1995.* Tunbridge Wells: South East Institute of Public Health. *(Equivalent volumes for Bromley, Bexley & Greenwich also produced)* [18]

Markus TA (1993) 'Cold, Condensation and Housing Poverty' in *Burridge R. & Ormandy D (Eds) Unhealthy Housing: Research, remedies and reform.* E&FN Spon, London. [7]

Markus, T (Ed.) (1994) *Domestic Energy and Affordable Warmth,* Watt Committee Report No.30, E&FN Spon, London. [10]

Martin CJ, Platt SD, Hunt SM (1987) Housing conditions and ill-health. *Brit. Med. J.* **294** 1125-1127. [5]

Martin CJ, Platt SD and Hunt SMI (1987) Housing conditions and ill-health. *Brit. Med. J.* **294**, 1125-27 [6, 11]

Maunsell K, Wraith DG, Cunnington AM (1968) Mites and house dust allergy in bronchial asthma. *The Lancet* **1** 1267-1270. [5]

McDowall M (1981) Long term trends in seasonal mortality. *Population Trends,* **26**, 16-19. [4]

McHorney CA, Ware JE, Raczek AK (1993) The MOS 36 item Short-Form Health Survey (SF-36): II. Psychometric and clinical tests of validity in measuring physical and mental health constructs. *Med Care* **31**: 247-63. [7]

McHorney CA, Ware JE, Lu JFR, *et al.* (1994) The MOS 36 item Short-Form Health Survey (SF-36): III. Tests of data quality and assumptions and reliability across diverse patient groups. *Med. Care* **32:** 40-52 [7]

McIntyre D (1993) Mites, asthma and fresh air. *Building Services* **March** 22-24. [5]

McIntyre D (1993) Mechanical Ventilation and house dust mites. *Clin Exp Allergy* **23(1)** 58 (abstr). [5]

McKee CM (1989) Deaths in Winter: Can Britain learn from Europe? *European J. of Epidemiology*, **5**, 2, 178-182. [2, 4]

Millqvist E, Bengtsson U, Bake B (1987) Occurrence of breathing problems induced by cold climate in asthmatics- a questionnaire survey. *Eur Respir J*, **71**: 444-9 [1]

Milne G and Boardman B (1997) *Making cold homes warmer: the effect of energy efficiency improvements in low-income homes.* Report to the Eaga Charitable Trust. Environmental Change Unit, University of Oxford. [10]

Mitchell EB, Wilkins S, Deighton J, Platts-Mills TAE (1985) Reduction of house dust mite allergen levels in the home: use of the acaricide Pirimiphos methyl. *Clin Allergy* **15** 235-240. [5]

Mittleman MA, Maclure M, Tofler GH, Sherwood JB, Goldberg RJ, Muller JE (1993). Triggering of acute myocardial infarction by heavy physical exertion. *New England J. of Medicine*, **329**, 1677-1683. [4]

Miyamoto T, Oshima S, Ishizaki T, Sato S (1968) Allergic identity between the common floor mite (Dermatophagoides Farinae) and house dust as a causative antigen in bronchial asthma. *J Allergy* **42** 14-28. [5]

Molyneux P and Palmer J (1997) *Towards a strategy for housing and health: Cost drivers and blocks that impact on the public's health.* London: Health and Housing. [18]

Morris JN, Everitt MG, Pollard R (1980) Vigorous exercise in leisure time: protection against coronary heart disease. *Lancet*, **2**, 1207-1210. [4]

Morrow Brown H, Merrett TG (1991) Effectiveness of an acaricide in management of house dust allergy. *Annals of Allergy* **67** 25-31. [5]

Mulla MS, Markrider JR, Galant SP, Amin L (1975) Some house dust control measures and abundance of *Dermatophagoides* mites in southern California. *J Med Entomol* **12** 5-9. [5]

Muller JE, Stone PH, Zuri ZG *et al.* (1985) Circadian variation in the frequency of onset of acute myocardial infarction. *New England J. of Medicine*, **313**, 1315-22 [4]

Muller JE, Tofler GH, Stone PH (1989) Circadian variation and triggers of onset of acute cardiovascular disease. *Circulation*, **79**, 733-743. [4]

Mundal R, Kjeldsen SE, Sandvik L, Erikssen G, Thaulow E and Erikssen J (1997) Seasonal co-variation in physical fitness and blood pressure at rest and during exercise in healthy middle aged men. *Blood Pressure*, **6**, 269-273. [4]

Murphy MFG and Campbell MJ (1987) Sudden Infant Death Syndrome and environmental temperatures: an analysis using vital statistics. *J. of Epidemiology and Community Health*, **41**, 63-71.

National Housing Federation (1997) *Housing for health*. London: National Housing Federation. [18]

NEA (1999) *Energy Action*, no. 77: p.6, National Energy Action, Newcastle upon Tyne. [10]

Neild PJ, Syndercombe-Court D, Keatinge WR, Donaldson GC, Mattock M, Caunce M (1994) Cold induced increases in erythrocyte count, plasma cholesterol and plasma fibrinogen of elderly people without a comparable rise in Protein C or Factor X. *Clinical Science*, **86**, 43-48. [1, 2, 4]

NES (1996) *National Home Energy Rating. Best Practice Guide to Energy Auditing*. National Energy Services Limited. Milton Keynes [12]

Ness AR, Powles JW, Khaw KT (1996) Vitamin C and cardiovascular disaease: a systematic review. *J Cardiovasc Risk* **3**:513-21 [2]

New Scientist (1999) p 49, 10 July 1999 [19]

NHS Executive (1996) *Burdens of Disease*, Department of Health, London [11]

Nicholson KG, Kent J, Hammersley V *et al.* (1997) Acute viral infections of upper respiratory tract in elderly people living in the community: comparative, prospective population based study of disease burden. *Brit. Med. J.*, **315**, 1060-4. [3]

Office for National Statistics (1996) Family Expenditure Survey 1995-96. *Family spending, Section 1.3 – Detailed household expenditure by gross income decile group.* Office for National Statistics, HMSO, London. [5]

ONS Office for National Statistics (1997) *Mortality statistics, cause 1996*. Series DH2 no 22 Table 3. The Stationery Office, London [11]

OPCS Office of Population, Censuses and Surveys (1987) *Trends in Respiratory Mortality 1951-1975*. Series DH1 No 7, 13-22. London: HMSO. [4]

Office of Population, Censuses and Surveys (1991) *Census of Population*. HMSO [15]

OPCS (1995) *Morbidity statistics from general practice*. Fourth national study 1991-1992 Series MB5 no 3, HMSO, London [11]

Offices, Shops and Railway Premises Act (1963) Her Majesty's Stationery Office, London. [3]

Omran M, Russell G (1996) Continuing increase in respiratory symptoms and atopy in Aberdeen schoolchildren. *Brit. Med. J.* **312** 34. [5]

Oreszczyn T and Pretlove S (1999) Condensation Targeter II, *Building Serv Eng Res & Technol*, **20** (4) [9]

Oreszczyn T (1992) Insulating the existing housing stock: mould growth and cold bridging, in *Energy Efficient Building (Eds: Roaf S & Hancock M)*, Blackwell Scientific Press [9]

Page D (1993) *Building Communities: A Study of New Housing Association Estates*, Joseph Rowntree Foundation, York [17]

Parker JD, Testa MA, Jimenez AH *et al.* (1994) Morning increase in ambulatory ischaemia in patients with stable coronary heart disease: importance of physical activity and increased cardiac demand. *Circulation,* **89**, 604-614. [4]

Pfaffenbarger RS Jr and Hyde RT (1984) Exercise in the prevention of coronary heart disease. *Preventative Medicine,* **13**, 3-22. [4]

Pfaffenbarger RS Jr, Hyde RT, Wing AL, Lee IM, Jung DL, Kampert JB (1993) The association of changes in physical activity level and other lifestyle characteristics with mortality among men. *New England J. of Medicine,* **328**, 538-545. [4]

PHA (1999) *External wall insulation to solid wall dwellings,* Penwith Housing Association, [19]

Pickering TG (1990) The clinical significance of diurnal blood pressure variations. *Circulation,* **81**, 700-702. [4]

Pickering TG, Gerin W, James GD, Pieper C, Schlussel YL and Schnall PL (1992) The effect of occupational and domestic stress on the diurnal rhythm of blood pressure (in) *Temporal Variations of the Cardiovascular System* (Eds: Schmidt, Engel and Blümchen), 272-282. Berlin: Springer-Verlag. [4]

Platt SD, Martin CJ, Hunt SM, Lewis CW (1989) Damp Housing, mould growth, and symptomatic health state. *Brit. Med. J.* **298** 1673-1678. [3, 5, 11, 12, 18]

Platt-Mills TAE (1994) How environment affects patients with allergic disease: indoor allergens and asthma. *Annals of Allergy* **72** pp381-3 [12]

Platts-Mills TAE, de Weck AL (1989) Dust mite allergens and asthma - a worldwide problem. *J Allergy Clin Immunol* **83** 416-427. [5]

Platts-Mills TAE, Hayden ML, Chapman MD, Wilkins SR (1987) Seasonal Variation in dust-mite and grass pollen allergens in dust from the houses of patients with asthma. *J Allergy Clin Immunol* **79** 781-791. [5]

Platts-Mills TAE, Thomas W, Aalberse RC, Vervloet D, Chapman MD (1992) Dust mite allergens and asthma: report of a second international workshop. *J Allergy Clin Immunol* **89** 1046-1060. [5]

Pollard K, Armstrong D and Bartholomew J (1997) *Housing and health: Assessing the impact of redevelopment on the health status of Liddle Ward.* London: United Medical and Dental Schools of Guy's and St Thomas'. [18]

Pollock JI and Golding J (1995) Social epidemiology of chickenpox in two British national cohorts. *J. of Epidemiology and Community health,* **47**, 274-81. [3]

Prescott-Clarke P and Primatesta P (1997) *Health Survey for England 1995. Volume I: Findings.* London: The Stationery Office. [18]

Prineas RJ, Gillum RF, Horibe H, Hannan PJ, Stat M (1980) The Minneapolis Children's Blood Pressure Study.   Standards of measurement for children's blood pressure. *Hypertension,* **2**, Suppl I: I16 - I18. [4]

Purcell HJ, Gibbs SR, Coats AJS and Fox KM (1992) Ambulatory blood pressure monitoring and circadian variation of cardiovascular disease; clinical and research applications. *International J. of Cardiology,* **36**, 135-149. [4]

Rasmussen, F, Borchsenius L, Winslow J *et al.* (1991) Associations between housing conditions, smoking habits and ventilatory lung function in men with clean jobs. *Scandinavian J. of Respiratory Diseases*, **59**, 264-76. [3]

Raw GJ and Hamilton RM (1995) *Building Regulation and Health*. Building Research Establishment, Garston, Watford. [3, 11]

Raw GJ, Whitehead C, Herbert JG and Cayless S (1996) *English house Conditions survey: analysis of health data,* Building Research Establishment, Watford [11]

Rayner KF (1994) 'Comfort and health: Cause for concern. *J. of the Royal Society of Health,* **114**: 153-6. [18]

Rees J, Price J (1993) *ABC of Asthma - Third Edition.* BMJ Publishing group, (49pp). [5]

Robertson T (1986) Circadian variation in the frequency of onset of stroke. *J. of American College of Cardiology*, 7, 2, 40A. [4]

Rona RJ, Chinn S, Burney PGJ (1995) Trends in the prevalence of asthma in Scottish and English primary school children 1982-92. *Thorax* 50 992-993. [5]

Rose G (1961) Seasonal Variation in Blood Pressure in Man. *Nature*, **189**, 4760, 235 [4]

Rose G (1966) Cold weather and ischaemic heart disease. *British J. of Preventative and Social Medicine*, **20**, 97-100. [4]

Rudge J (1995) Why does it appear that the British have traditionally cared so little for energy efficiency and warmth in the home? *Proceedings CIBSE 1995 National Conference*, Chartered Institution of Building Services Engineers, London. [10]

Rudge J (1996) 'British weather: conversation topic or serious health risk?' *International Journal of Biometeorology* 39: 151-155 [10]

Russell DW, Fernandez-Caldas E, Swanson MC, Seleznick MJ, Trudeau WL, Lockey RF (1991) Caffiene, a naturally occuring acaricide. *J Allergy Clin Immunol* 87 107-110. [5]

Saez M, Sunyer J, Castellsague J, Murillo C, Anto JM (1995) Relationship between weather temperature and mortality: A time series analysis approach in Barcelona. *Int J Epidemiol* **24(3)**:576-582 [2]

Sandvik L, Erikssen J, Thaulow E, Erikssen G, Mundal R and Rodahl K (1993) Physical fitness as a predictor of mortality among healthy, middle-aged Norwegian men. *New England J. of Medicine*, **328**, 533-537. [4]

SCOPH (1994) *Housing, homelessness and health*. The Standing Conference on Public Health working group report, Nuffield Provincial Hospitals Trust, London [11]

Scottish Consumer Council (1978) *Houses to mend. A survey on council house repairs in Scotland.* Scottish Consumer Council, Glasgow. [5]

Scott-Samuel A, Birley M, Ardern K (1998) *The Merseyside Guidelines for Health Impact Assessment.* Liverpool:The University of Liverpool [18]

Seedhouse, D (1986) *Health, The Foundation for Achievement*, John Wiley, Chichester [17]

Sesay HR, Dobson RM (1972) Studies on the mite fauna of house dust in Scotland with special reference to that of bedding. *Acarologia* 14 384. [5]

Shapiro D and Goldstein IB (1998) Wrist actigraph measures of physical activity level and ambulatory blood pressure in healthy elderly persons. *Psychophysiology*, **35**, 305-312. [4]

Shapiro S, Skinner E, Karemer M *et al.* (1985) 'Measuring need for mental health services in a general population'. *Medical Care,* **23**: 1033-43. [18]

SHCS (1996) *Scottish House Condition Survey 1996* Scottish Homes, Edinburgh 1997 [5, 12]

Sloan DSG (1996) Number of excess deaths during winter is large. Letter. *Brit. Med. J.* **312**, 24 [11]

Sloan REG and Keatinge WR (1975) Depression of sublingual temperature by cold saliva. *Brit. Med. J.,* **1**, 718-721. [1]

Sly RM, Josephs SH, Eby DM (1985) Dissemination of dust by central and portable vacuum cleaners. *Annals of Allergy* **54** 209-212. [5]

Smith SJ, Alexander A and Hill S (1993) *Housing provision for people with health and mobility needs: A guide to good practice.* York: Joseph Rowntree Foundation. [18]

Social Exclusion Unit (1998) *Bringing Britain together: A national strategy for neighbourhood renewal.* Cm 4045. London: The Stationery Office. [13, 18]

Somerville M, Mackenzie I, Owen P, Miles D (1999) Housing and health. Does installing heating in their houses improve the health of children with asthma? *(in preparation)* [12]

Souhrada, M and Souhrada, JF (1981) The direct effect of temperature on airway smooth muscle. *Respiratory Physiology*, **44**, 311-23. [3]

Spieksma FThM, Spieksma-Boezeman MIA (1967) The mite fauna of house dust with particular reference to the house-dust mite *Dermatophagoides Pteronyssinus* (Troussart 1897). *Acarologia* **11** 226-241. [5]

Sporik R, Chapman MD, Platts-Mills TAE (1992) House dust mite exposure as a cause of asthma. *Clin Exp Allergy* **22** 897-906. [5]

Standing Conference on Public Health (1994) *Housing, homelessness and health.* London: Nuffield Provincial Hospitals Trust. [18]

Steen N, Hutchinson A, McColl E, Eccles MP, Hewison J, Meadows KA, Blades SM, Fowler P (1994) Development of a symptom-based outcome measure for asthma. *Brit. Med. J.* **309** pp1065-8 [12]

Stephens NL, Cardinal R and Simmons B (1977) Mechanical properties of tracheal smooth muscle: effect of temperature. *American J. of Physiology*, **233**, C 92-8. [3]

Stewart AL, Hays RD, Ware JE (1988) The MOS 36 item Short-Form Health Survey: II. Reliability and validity in a patient opulation. *Med Care* **26**(7): 724-35 [7]

Stout RW and Crawford V (1991) Seasonal variations fibrinogen concentrations among elderly people. *Lancet*, **338**, 9-13. [4]

Stout RW, Crawford VLS, Scarabin PY, Bara L, Nicaud V, Cambou JP, Arveiler D, Luc G, Evans AE, Cambien F (1994) Seasonal variations of plasma fibrinogen in elderly people (8). *Lancet* **343**(8903):975-976 letter [2]

Strachan DP (1988) Damp housing and childhood asthma: validation of reporting of symptoms. *Brit. Med. J.* **297** pp1223-6 [5, 12]

Strachan DP (1993) Dampness, mould growth and respiratory disease in children, in *Unhealthy Housing: Research, Remedies and Reform (Eds. Burridge R and Ormandy D)*, E & FN Spon, London, pp 94-116. [3, 7]

Strachan DP, Anderson HR, Limb ES, O'Neill A, Wells N (1994) A national survey of asthma prevalence, severity and treatment in Great Britain. *Arch Dis Child* **70** 174-178. [5]

Strachan DP, Carey IM (1995) Home environment and severe asthma in adolescence: a population based case-control study. *Brit. Med. J.* **311** pp1053-6 [12]

Strachan DP, Sanders CH (1989) Damp housing and childhood asthma: respiratory effects of indoor temperature and relative humidity. *J Epidemiology Community Health* **43** 7-14. [5, 12]

Tenant's Resource Information Service (1995) *Action in Damp Homes* London [6]

Tillett HE, Smith JWG and Gooch CD (1983) Excess deaths attributable to influenza in England and Wales: age at death and certified cause. *International J. of Epidemiology*, **12**, 344-52. [3]

Tofler GH, Brezinski D, Schafer A *et al.* (1987) Concurrent morning increase in platelet aggregability and the risk of myocardial infarction and sudden cardiac death. *New England J. of Medicine*, **316**, 1514-1518. [4]

Toth, LA and Blatteis, CM (1995) Adaptation to the microbial environment, in *Handbook of Physiology*, Section 2 (Ed. MJ Fregly), Oxford University Press, New York, 1489-1519. [3]

Tovey ER, Chapman MD, Platts-Mills TAE (1981) Mite faeces are a major source of house dust allergens. *Nature* **289** 592-593. [5]

Tovey ER, Chapman MD, Wells CW, Platts-Mills TAE (1981) The distribution of dust mite allergen in the houses of patients with asthma. *Am Rev Repsir Dis* **124** 630-635. [5]

Truro (1896) *Medical Officer's Report*, City of Truro, Truro, Cornwall [12]

Tsementzis SA, Gill JS, Hitchcock ER, Gill SK, Beevers DG (1985) Diurnal variation of and activity during the onset of stroke. *Neurosurgery*, **17**, 901-904. [4]

Turos M (1979) Mites in house dust in the Stockholm area. *Allergy* **34** 11-18. [5]

Tyrrell DAJ (1965) *Common Colds and Related Diseases.* Edward Arnold, London. [3]

Uitenbroek DG (1993) Seasonal variation in leisure time physical activity. *Medicine and Science in Sports and Exercise*, **25**, 6, 755-760. [4]

Urban Task Force (1999) *Towards an Urban Renaissance,* Final Report of the Urban Task Force, E & FN Spon, [19]

van Bronswijk JEMH and Sinha RH (1971) Pyroglyphid mites (acari) and house dust allergy. *J Allergy* **47** 31-52. [5]

van Bronswijk JEMH (1973) Dermatophapoides Pteronyssinus (Troussart 1897) in mattress and floor dust in a temperate climate (Acari: Pyroglyphidae). *J Med Ent* **1** 63-70. [5]

van Bronswijk JEMH (1973) The effectiveness of a vacuum cleaner in removing dust , mites and possibly mite allergens from a carpet. *Airways* **4(2)** 10-16. [5]

van Egeren LF (1992) Effects of behavioural rhythms on blood pressure rhythms (in) *Temporal Variations of the Cardiovascular System* (Eds: Schmidt, Engel and Blümchen). Berlin: Springer-Verlag, 283-296. [4]

Verhoeff AP, van Strien RT, van Wijnen JR Brunekreef B (1995) Damp housing and childhood respiratory symptoms: the role sensitisation to dust mites and molds. *Amer. J. Epidemiology* **141** pp103-10 [12]

Voorhorst R, Spieksma FTH, Varekamp H, Leupen MJ, Lyklema AW (1967) The house dust mite (Dermatophagoides Pteronyssinus) and the allergens it produces: Identity with the house dust allergen. *J Allergy* **39** 325. [5]

Voorhorst R, Spieksma FTH, Varekamp H (1969) *House Dust Atopy and the House Dust Mite Dermatophagoides Pteronyssinus.* Stafleu's Scientific Publishing Company, Leiden. [5]

Voorips LE, Ravelli ACJ, Dongelmans PCA, Deurenberg P and van Staveren WA (1991) A physical activity questionnaire for the Elderly. *Medicine and Science in Sports and Exercise*, **23**, 8, 974-979. [4]

Wade DT (1994) 'Stroke (acute cerebrovascular disease)' in Stevens A and Rafferty J (Eds). *Health Care Needs Assessment Volume I.* Oxford: Radcliffe Medical Press. [18]

Wadhams (1998) *Upwardly Mobile,* Chris Wadhams Associates [17]

Walshaw MJ and Evans CC (1986) Allergen Avoidance in house dust mite sensitive adult asthma. *Quarterly J. of medicine* **58** 199 – 215. [5]

Ware JE and Sherbourne CD (1992) The MOS 36 item Short-Form Health Survey (SF-36):1. Conceptual framework and item selection. *Med Care* **30**: 473-83 [7]

Warner JA, Marchant JL, Warner JO (1993) Double blind trial of ionisers on children with asthma sensitive to the house dust mite. *Thorax* **48** 330 – 333. [5]

Warner JO, Price JA (1990) Aero-allergen avoidance in the prevention and treatment of asthma. *Clin Exp Allergy* **20(3)** 15-19. [5]

Watts AJ (1972) Hypothermia in the aged: a study of the role of cold sensitivity. *Environmental Research*, **5**, 119-126. [4]

West RR (1989) Seasonal Variation in CHD Mortality. *International J. of Epidemiology*, **18(2)** 463-464. [2, 4]

West RR and Lowe CR (1976) Mortality from ischaemic heart disease - inter-town variation and its association with climate in England and Wales. *International J. of Epidemiology*, **5**, 195-201. [4]

Wharton GW (1976) House dust mites. *J Med Ent* **12** 577-621.

Whitehead, C (1998) *Hackney Housing Conditions: Implications for Policy,* Hackney Housing Forum, London [17]

Whitrow D and Pycock R (Ed.) (1993) *House Dust Mites: How they affect asthma, eczema and other allergies.* Elliot Right Way Books, Surrey, England. [5]

Wilkinson P, Stevenson S, Armstrong B, Fletcher T (1998) Housing and winter death. *Epidemiology* **9(4)**:S59 [2]

Williamson IJ, Martin CJ, McGill G, Monic RDR Fennerty AG (1997) Damp housing and asthma: a case-control study. *Thorax* **52** pp229-234 [5, 12]

Willich SN (1997) *Circadian Variation and Triggering of Acute Cardiac Coronary Heart Disease.* Monheim, FRG: Schwarz. [4]

Willich SN, Goldberg RJ, MacClure M, Periello L, Muller JE (1992) Increased onset of sudden cardiac death in the first three hours after wakening. *American J. of Cardiology*, **70**, 65-68. [4]

Wilmshurst P (1994) Temperature and cardiovascular mortality: Excess deaths from heart disease and stroke in northern Europe are due in part to the cold. *Brit. Med J* **309**:1029-1030 [2]

Wilson PWF, Paffenbarger RS Jr, Morris JN (1986) Assessment methods for physical activity and physical fitness in population studies: report of a NHLBI workshop. *American Heart J.*, **111**, 1177-1192. [4]

Winnicki M, Canali C, Accurso V, Dorigatti F, Palatini P (1996) Relation of 24-hour ambulatory blood pressure and short-term blood pressure variability to seasonal changes in environmental temperature in stage I hypertensive subjects. *Clinical and Experimental Hypertension*, **18**, 8, 995-1012. [4]

Woodhouse PR, Khaw KT, Plummer M (1993) Seasonal variation of blood pressure and its relationship to ambient temperature in an elderly population. *J of Hypertension* **11(11)**:1267-1274 [2, 4]

Woodhouse PR, Khaw K-T, Plummer M, Foley E, Meade TW (1994) Seasonal variations in plasma fibrinogen and factor VII activity in the elderly: winter infections and death from cardiovascular disease. *Lancet,* **343**: 435-9. [1]

Woodhouse PR, Khaw KT, Plummer M *et al.* (1994) Seasonal variations of plasma fibrinogen and Factor VII in the elderly: winter infections and death from cardiovascular disease. *The Lancet*, **343**, 435-9. [3]

Woodin Consultancies (1996) *Just What the Doctor Ordered: Housing, Health and Community Safety on Holly Street,* Woodin Consultancies, [17]

Wright G, Payling and Wright H (1945) Etiological factors in bronchopneumonia among infants in London. *J. of Hygiene*, Cambridge, **44**, 15-30. [3]

Keyword index

Numbers refer to chapters